José J. Pazos Arias, Ana Fernández Vilas, and Rebeca P. Díaz Redondo

Recommender Systems for the Social Web

Intelligent Systems Reference Library, Volume 32

Editors-in-Chief

Prof. Janusz Kacprzyk
Systems Research Institute
Polish Academy of Sciences
ul. Newelska 6
01-447 Warsaw
Poland
E-mail: kacprzyk@ibspan.waw.pl

Prof. Lakhmi C. Jain
University of South Australia
Adelaide
Mawson Lakes Campus
South Australia 5095
Australia
E-mail: Lakhmi.jain@unisa.edu.au

Further volumes of this series can be found on our homepage: springer.com

Vol. 8. Edward H.Y. Lim, James N.K. Liu, and Raymond S.T. Lee
Knowledge Seeker – Ontology Modelling for Information Search and Management, 2011
ISBN 978-3-642-17915-0

Vol. 9. Menahem Friedman and Abraham Kandel
Calculus Light, 2011
ISBN 978-3-642-17847-4

Vol. 10. Andreas Tolk and Lakhmi C. Jain
Intelligence-Based Systems Engineering, 2011
ISBN 978-3-642-17930-3

Vol. 11. Samuli Niiranen and Andre Ribeiro (Eds.)
Information Processing and Biological Systems, 2011
ISBN 978-3-642-19620-1

Vol. 12. Florin Gorunescu
Data Mining, 2011
ISBN 978-3-642-19720-8

Vol. 13. Witold Pedrycz and Shyi-Ming Chen (Eds.)
Granular Computing and Intelligent Systems, 2011
ISBN 978-3-642-19819-9

Vol. 14. George A. Anastassiou and Oktay Duman
Towards Intelligent Modeling: Statistical Approximation Theory, 2011
ISBN 978-3-642-19825-0

Vol. 15. Antonino Freno and Edmondo Trentin
Hybrid Random Fields, 2011
ISBN 978-3-642-20307-7

Vol. 16. Alexiei Dingli
Knowledge Annotation: Making Implicit Knowledge Explicit, 2011
ISBN 978-3-642-20322-0

Vol. 17. Crina Grosan and Ajith Abraham
Intelligent Systems, 2011
ISBN 978-3-642-21003-7

Vol. 18. Achim Zielesny
From Curve Fitting to Machine Learning, 2011
ISBN 978-3-642-21279-6

Vol. 19. George A. Anastassiou
Intelligent Systems: Approximation by Artificial Neural Networks, 2011
ISBN 978-3-642-21430-1

Vol. 20. Lech Polkowski
Approximate Reasoning by Parts, 2011
ISBN 978-3-642-22278-8

Vol. 21. Igor Chikalov
Average Time Complexity of Decision Trees, 2011
ISBN 978-3-642-22660-1

Vol. 22. Przemysław Różewski, Emma Kusztina, Ryszard Tadeusiewicz, and Oleg Zaikin
Intelligent Open Learning Systems, 2011
ISBN 978-3-642-22666-3

Vol. 23. Dawn E. Holmes and Lakhmi C. Jain (Eds.)
Data Mining: Foundations and Intelligent Paradigms, 2011
ISBN 978-3-642-23165-0

Vol. 24. Dawn E. Holmes and Lakhmi C. Jain (Eds.)
Data Mining: Foundations and Intelligent Paradigms, 2011
ISBN 978-3-642-23240-4

Vol. 25. Dawn E. Holmes and Lakhmi C. Jain (Eds.)
Data Mining: Foundations and Intelligent Paradigms, 2011
ISBN 978-3-642-23150-6

Vol. 26. Tauseef Gulrez and Aboul Ella Hassanien (Eds.)
Advances in Robotics and Virtual Reality, 2011
ISBN 978-3-642-23362-3

Vol. 27. Cristina Urdiales
Collaborative Assistive Robot for Mobility Enhancement (CARMEN), 2011
ISBN 978-3-642-24901-3

Vol. 28. Tatiana Valentine Guy, Miroslav Kárný and David H. Wolpert (Eds.)
Decision Making with Imperfect Decision Makers, 2012
ISBN 978-3-642-24646-3

Vol. 29. Roumen Kountchev and Kazumi Nakamatsu (Eds.)
Advances in Reasoning-Based Image Processing Intelligent Systems, 2012
ISBN 978-3-642-24692-0

Vol. 30. Marina V. Sokolova and Antonio Fernández-Caballero
Decision Making in Complex Systems, 2012
ISBN 978-3-642-25543-4

Vol. 31. Ludomir M. Laudański
Between Certainty and Uncertainty, 2012
ISBN 978-3-642-25696-7

Vol. 32. José J. Pazos Arias, Ana Fernández Vilas, and Rebeca P. Díaz Redondo
Recommender Systems for the Social Web, 2012
ISBN 978-3-642-25693-6

José J. Pazos Arias, Ana Fernández Vilas,
and Rebeca P. Díaz Redondo

Recommender Systems for the Social Web

Dr. José J. Pazos Arias
E.T.S.E. de Telecomunicación
Campus Universitario de Vigo
36310 Vigo
Spain
E-mail: jose@det.uvigo.es

Dr. Rebeca P. Díaz Redondo
E.T.S.E. de Telecomunicación
Campus Universitario de Vigo
36310 Vigo
Spain
E-mail: rebeca@det.uvigo.es

Dr. Ana Fernández Vilas
E.T.S.E. de Telecomunicación
Campus Universitario de Vigo
36310 Vigo
Spain
E-mail: avilas@det.uvigo.es

ISBN 978-3-642-44627-6
ISBN 978-3-642-25694-3 (eBook)

DOI 10.1007/978-3-642-25694-3

Intelligent Systems Reference Library
ISSN 1868-4394

Typeset by Scientific Publishing Services Pvt. Ltd., Chennai, India.

Printed on acid-free paper

9 8 7 6 5 4 3 2 1

springer.com

Preface

The emergence and deployment of new digital communication technologies (3G mobile networks, digital television, xDSL, home automation networks, etc.) has led to a gradual increase in speed while a steady decline in prices at Internet connections. Moreover, since much longer ago, the continuing advances in electronic technology have led to a steady rise in computer performance at a price increasingly smaller, which has turned the computer into a product accessible to everyone, at least in developed countries. More recently, again advances in electronic technology have made possible the emergence of new Internet devices (game consoles, mobile phones, PDAs, tablets, etc.) that paint a scenario in which ubiquitous access to the Information Society is no longer an illusion.

In this new scenario, full of unquestionable strengths and opportunities, one would expect that users would benefit from the potential of ubiquitous access (from home or from anywhere via mobile devices) to an increasing amount of contents and interactive services. We talk about a new technological playground which only seems be restricted by the capacity for innovation and economic viability of this innovation; a huge space of possibilities which encompass any imaginable and even unimaginable application domain, such as entertainment, news, electronic commerce, learning, games, multimedia, ... In short, every conceivable e-technology; e-learning, e-administration, e-health, e-tourism, e-governance and so on.

Although certainly the offer of contents and services is increasing dramatically, the reality on the use of these new technologies is really quite different from what would be expected and can be described as a situation of information overload. Often users have within their reach so much information that finding what they need becomes a very tedious task, causing some disappointment, if not rejection to the use of these new technologies. In the early days, the problem of access to Internet information was that users had to know exactly what they wanted, where it was, and go directly there. From those days to present content search engines, we have come a long way. When the access device is a personal computer, the successful solution has its origin in the 90's with the development of search engines (Yahoo, Google, Bing, etc.). These systems implement different types of software algorithms in order to locate resources on the Internet from a textual (keyword-based) description, provided by the user. But even today, users are faced with situations where they have to narrow the search criterion again and again to avoid ending up lost among thousands or millions of results, and really find what they seek. On the other hand, information overload can also lead to negative effects on the provider side. It can produce a significant reduction in the expected benefits as content and services are diluted in a vast and disorganized mass of content and, therefore, may go undetected for many users

Despite their widely recognized shortcomings, the success of search engines is undeniable and it is confirmed by everyday use which can be seen in users accessing the Internet from PCs. Unfortunately, the usage model that is implemented by these search engines is not exportable to many of the new access devices, nor to the universe of users that are intended to appeal to the use of Internet. On the one hand, the challenges stem from the lack of appropriate interfaces for text input in most of these new devices (game consoles, set-top-boxes, mobile phones, etc.), devices that are mainly handled with remote controls, touch screens and so on. On the other hand, many of these new devices are linked to usage habits radically different from the ones we're used to seeing on the PC users. This, coupled with the lack of technological background of many users, makes it unreasonable to expect too much initiative from these users, and even less reasonable to assume that users will be able to describe using keywords those resources that are of their interest.

In an effort to overcome the shortcomings of search engines aforementioned, in the late 90's emerged the so called recommender systems, which implement methods and algorithms developed in the field of artificial intelligence to automatically select both proactive and personalized resources that best suit the preferences and needs of each user. Consequently, recommender systems and search engines came up with the same goal, but its scope different, which actually makes them complementary.

On the other hand, although with different approaches and algorithms, the different search engines base their behavior on comparing strings, i. e. between the resources description given by the user and the metadata of these resources. Because of that, the quality of this solution often resents, which is clear when an overwhelming amount of resources is returned as searching result; what is more, many of them are slightly related with the original query. With the aim of overcoming this limitation, significant research efforts have been achieved in a relatively new field: the so-called semantic web or Web 3.0. The common issue is trying to provide the resources' descriptions of a significant meaning (semantics) to increase the precision of the search engines.

Regardless of which is the selected solution (search engines or recommender systems) the precision of the obtained results is strongly linked to the precision of the metadata that describe each available resource. In fact, all standard metadata specification relies on the resources' creators or providers, who often use a controlled mechanism of classification and tagging. Consequently, using this strict model is currently being questioned because of both practical and social issues. The former because the metadata linked to available resources, if any, are usually vague and imprecise. This is directly caused by the impossibility of doing this characterization of a constantly growing volume of products by hand, and the current automated characterization systems (based on voice or images recognition, texts analysis, etc.) are still in a very immature stage of development. The latter, social aspects, because the metadata specification standards, based on controlled vocabulary previously fixed, impose a very rigid system, which constitutes a barrier to encourage the participation of typical users, who usually are unacquainted with this standards and rules. Besides, this strict scheme entails that resources' descriptions are biased by the creator and provider's perspective, totally ignoring the consumers' point of view. The successful implementation of the so-called Web 2.0 technologies gives a more important role for users, who actually act not only as typical consumers, but also as producers of information so they cannot be pushed aside any longer.

Involving users in the tasks of giving information of available resources requires, on the one hand, getting away from formal metadata specification and, on the other hand, encouraging them to have fully cooperation. Using natural language for resources' description entails dealing with huge challenges to set up recommending systems or browsers, since it will be essential to modify the current intelligent information processing systems based on rigid structure (ontologies) to tackle new and flexible knowledge structures (folksonomies). Ambiguity is an inherent quality of folksonomies which involves a less effective reasoning than using ontologies. However, everything point out that the best solution is not a pure solution, but a hybrid one that would be able to exploit their complementarities.

Having mentioned this background, this book has collected contributions from well-known experts from academy and industry to provide a broader view of the Recommender Systems for the Social Web field. The book aims to help readers to discover and understand the previously mentioned objectives, the functionality that underlie and the main challenges that must be undertaken.

Vigo, 29th September 2011 *The Editors*

Acknowledgements

The editors wish to acknowledge the contribution of all the authors in clarifying the field of Social Recommendations, surely helping researchs and academia to foster the advance in the new Knowledge Society.

Also thanks to the financial support of private and public institutions which have facilitated the editors' know-how in this topic. Specially to *Ministerio de Educacion y Ciencia (Gobierno de Espana)* research project TIN2010-20797 (partly financed with FEDER funds), and to *Conselleria de Educacion e Ordenacion Universitaria (Xunta de Galicia)* incentives file CN 2011/023 (partly financed with FEDER funds).

Contents

List of Contributors

Stefan Birnkammerer
Faculty of Informatics, Technische Universität München, Boltzmannstr 3, 85748 Garching, Germany
e-mail: birnkammerer@in.tum.de

Celeste Campo
Department of Telematic Engineering, University Carlos III of Madrid. 28911 Leganés, Madrid, Spain
e-mail: celeste@it.uc3m.es

Iván Cantador
Universidad Autónoma de Madrid, Madrid, Spain
e-mail: ivan.cantador@uam.es

Pablo Castells
Universidad Autónoma de Madrid, Madrid, Spain
e-mail: pablo.castells@uam.es

Li Chen
Department of Computer Science, Hong Kong Baptist University, Hong Kong
e-mail: lichen@comp.hkbu.edu.hk

Rebeca P. Díaz Redondo
Department of Telematic Engineering. University of Vigo
e-mail: rebeca@det.uvigo.es

Ana Fernández Vilas
Department of Telematic Engineering. University of Vigo
e-mail: avilas@det.uvigo.es

Carlos García-Rubio
Department of Telematic Engineering, University Carlos III of Madrid. 28911 Leganés, Madrid, Spain
e-mail: cgr@it.uc3m.es

Alberto Gil-Solla
Department of Telematic Engineering. University of Vigo
e-mail: agil@det.uvigo.es

Georg Groh
Faculty of Informatics, Technische Universität München, Boltzmannstr 3,
85748 Garching, Germany
e-mail: grohg@in.tum.de

Ido Guy
IBM Research-Haifa, 31905, Israel
e-mail: ido@il.ibm.com

Andreas Hotho
Data Mining and Information Retrieval Group at LS VI, University of Würzburg,
Am Hubland, 97074 Würzburg, Germany
e-mail: hotho@informatik.uni-wuerzburg.de

Robert Jäschke
Knowledge & Data Engineering Group, University of Kassel,
Wilhelmshöher Allee 73, 34121 Kassel, Germany
e-mail: jaeschke@cs.uni-kassel.de

Valeria Köllhofer
Faculty of Informatics, Technische Universität München, Boltzmannstr 3,
85748 Garching, Germany
e-mail: koellhof@in.tum.de

Manuela I. Martín-Vicente
Department of Telematic Engineering. University of Vigo
e-mail: mvicente@det.uvigo.es

Marcos Martínez-Romero
IMEDIR Center, University of A Coruña, 15071 A Coruña, Spain
e-mail: marcosmartinezp@udc.es

Folke Mitzlaff
Knowledge & Data Engineering Group, University of Kassel,
Wilhelmshöher Allee 73, 34121 Kassel, Germany
e-mail: mitzlaff@cs.uni-kassel.de

Alejandro Pazos
Department of Information and Communication Technologies,
Computer Science Faculty, University of A Coruña, 15071 A Coruña, Spain
e-mail: apazos@udc.es

José J. Pazos Arias
Department of Telematic Engineering. University of Vigo
e-mail: jose@det.uvigo.es

Javier Pereira
IMEDIR Center, University of A Coruña, 15071 A Coruña, Spain
e-mail: javierp@udc.es

Manuel Ramos-Cabrer
Department of Telematic Engineering. University of Vigo
e-mail: mramos@det.uvigo.es

Alicia Rodríguez-Carrión
Department of Telematic Engineering, University Carlos III of Madrid. 28911 Leganés, Madrid, Spain
e-mail: arcarrio@it.uc3m.es

Teresa Rodríguez de las Heras Ballell
University Carlos III of Madrid
e-mail: teresa.rodriguezdelasheras@uc3m.es

Gerd Stumme
Knowledge & Data Engineering Group, University of Kassel, Wilhelmshöher Allee 73, 34121 Kassel, Germany
e-mail: stumme@cs.uni-kassel.de

José M. Vázquez-Naya
IMEDIR Center, University of A Coruña, 15071 A Coruña, Spain
e-mail: jmvazquez@udc.es

Quan Yuan
IBM Research - China, Zhongguancun Software Park, Haidian District, Beijing, China
e-mail: quanyuan@cn.ibm.com

Shiwan Zhao
IBM Research - China, Zhongguancun Software Park, Haidian District, Beijing, China
e-mail: zhaosw@cn.ibm.com

Part I
General Aspects

Chapter 1
Social Recommender Systems

Georg Groh, Stefan Birnkammerer, and Valeria Köllhofer

Abstract. In this contribution we review and discuss limits and chances of social recommender systems. After classifying and positioning social recommender systems in the basic landscape of recommender systems in general via a short review and comparison, we present related work in this more specialized area. After having laid out the basic conceptual grounds, we then contrast an earlier study with a recent study in order to investigate the limits of applicability of social recommenders. The earlier study replaces rating-similarity-based neighbourhoods in collaborative filtering with subgraphs of the user's social network (social filtering) and investigates the performance of the resulting classifier in a taste related domain. The other study which is discussed in more detail investigates the applicability of the method to recommendations of more factual, content-oriented items: posts in discussion boards. While the former study showed that the social filtering approach works very well in taste related domains, the second study shows that a mere transplantation of the idea to a more factual domain and a situation with sparse social network data does perform less satisfactorially.

Keywords: Social Recommender Systems, Social Context, Collaborative Filtering.

1.1 Introduction

Recommender systems, in general, aim at supporting users by recommending them previously unseen items from a (in most cases homogeneous) set of such items. Items typically encompass products, services, media items (films, music etc.), information items (as in news filtering systems), collections of information items (webpages, portals etc.).

Georg Groh · Stefan Birnkammerer · Valeria Köllhofer
Faculty of Informatics, Technische Universität München, Boltzmannstr 3,
85748 Garching, Germany
e-mail: {grohg,koellhof,birnkammerer}@in.tum.de

J.J. Pazos Arias et al.: Recommender Systems for the Social Web, ISRL 32, pp. 3–42.
springerlink.com

One of the best known recommender system may be the recommender of Amazon [4]. However there are many other recommender systems on the Web, e.g. classic movie recommenders such as MovieLens (http://www.movielens.org) or music recommenders as Ping of Itunes (http://www.apple.com/itunes) or LastFM (http://www.lastfm.de).

In general, recommender systems are usually subdivided into two main streams: *content based* and *collaborative based*, which will be discussed in the following sections. Hybrid approaches also exist which combine content based and collaborative techniques. Content based recommender focus on the current user's rating history and similarity between items to compute recommendations while classic collaborative recommender systems use implicit or explicit ratings by other users and comparisons of rating history of the current user with other users to provide meaningful recommendations for the current user. A recommender can e.g. either predict a rating a user would make for a given object (prediction problem) or use these predictions to present the best N items as recommendations (top-N recommendation problem). A variant of collaborative recommenders are Social Recommender systems which we will discuss in more depth in this contribution.

In Social Recommender systems, aspects and components of traditional recommenders are explicitly designed using social entities. The concept may include the case where recommended "items" may themselves be social entities such as persons or groups of persons (e.g. in expert-recommenders). However, the interpretation we will mainly focus on centers on the concept of substituting the user-neighborhood, whose ratings or attitude towards the item-set are considered to be similar to the current user's tastes or needs, by a social context. This social context may be less dynamic (e.g. the friends or another sub-network of a social network in a Social Networking platform (such as Facebook) or short term (e.g. the social situation a user is in (see e.g. [34]).

In the following sections we will discuss aspects of social recommender systems in more detail. In order to define common vocabulary, we will first review basic related work on recommender systems in general, shortly discussing content-based, collaborative filtering based and other recommenders. We then review related work on social recommender systems. After that, we shortly review an earlier study, comparing classic collaborative filtering with social filtering in a taste related domain, in order to lay the ground for a comparison with the results of a second study we then discuss. This study also investigates collaborative filtering vs. social filtering in a domain which is more information centric. Both studies are compared and discussed. We conclude with a short summary and an outlook on future prospects.

1.2 Related Work on Recommender Systems

1.2.1 Content-Based Filtering

Content based recommenders make their predictions on the basis of utility $u(c,s)$ of item s for user c. The utility function u is constructed from utility values $u(c,s_i)$ (e.g. ratings) that user c assigned to those items s_i which are similar to item s [2]. That is, prediction whether an item is useful for a user or not is made on the basis of the set S of items s_i that are similar to s and uses an aggregation of the ratings $u(c,s_i)$ of these items s_i to make a prediction for $u(c,s)$. This approach requires a formalization of similarity $\mathrm{sim}(s,s_i)$ of two items.

Because content-based filtering has its origin in information retrieval [5, 6, 2] and information filtering [7, 2] it often recommends items based on textual information [2]. The content of these textual items is e.g. described by keyword extraction and formalizing the significance of keywords via weights. There are several techniques to determine the weights of keywords in a document, e.g. term frequency/inverse document frequency (TF/IDF) [5, 2]. A typical example for a class of methods analogously used by content-based filtering are spam-filter approaches for e-mail, using text-classification methods (e.g. based on support vector machines [42]).

Problems

Content-based filtering (see e.g. [2, 56]) is a very good approach for recommending items such as textual information where similarity measures based on their immediate data-representation are available. It is difficult for items for which a semantically meaningful representation-based similarity measure is hard to construct (e.g. multimedia items). In this case, properties need to be assigned manually to items via meta-data. This point of *limited content-analysis* is one of the disadvantages of content-based filtering approaches.

Another issue of content based filtering is *overspecialization*, as in all approaches which are basically classification based. The effect of *overspecialization* is that a user may get recommendations for items being too similar to those he already knows *(portfolio effect)*. To overcome this issue, one can introduce randomness or can filter out items that are too similar [9, 71, 2]. If a user is new to the system no (accurate) recommendations could be made for him unless he has assembled and rated an adequate number of items *(New User Problem)*. This challenge is present in many recommender approaches. Also, if the number of items initially present in the system which can be considered for recommendation is too small, the recommendations may be of limited value which is one flavor of a *Cold Start Problem*, which is also common to other recommender approaches.

1.2.2 Collaborative Filtering

In contrast to content-based filtering, collaborative filtering (CF) does not focus on the *content* of the concerning objects but on the opinion other uses have about this item. In essence, these algorithms work on sets of ratings, i.e. all ratings r for an item s provided by those users similar to user c are taken into account to make a prediction of this item for user c. The basic assumption behind this approach is, that people with similar taste will rate objects similar. There are different ways to categorize collaborative filtering approaches. [13] and [2] divide them into memory based (or heuristic-based) and model based algorithms, whereas [65] divides them into probability-based and non-probability-based.

Non-Probabilistic Approaches

User-based nearest neighbor algorithms predict a rating $r(s,c)$ for a user c and a previously unseen (unrated) item s with the help of ratings $\{r(s,c_{\text{sim}})\}$ by those users $\{c_{\text{sim}}\}$ who are most similar to the current user c. To overcome the bias introduced by a different general rating behavior of users the ratings of a user are considered relative to the average rating of that user. Some algorithms try to improve accuracy and efficiency by considering only the k nearest users [36] (the neighborhood). The prediction is usually computed as a weighted sum over $\{r(s,c_{\text{sim}})\}$ were user-user rating similarity may be used as the weights.

There are different approaches to define the similarity of two users with respect to their ratings. [60] proposed to use *Pearson-Correlation*, whereas [63] and [13] proposed cosine-similarity. [13] also proposed some modification to improve performance of the algorithm: *default voting, inverse user frequency, case amplification* and *weighted majority prediction*. The user-based nearest neighbor algorithm will be described in detail in Section 1.5.

Object-based nearest neighbor algorithms are an alternative approach proposed by [63]. They use similarity between items, but not in the exact sense of content-based filtering. For a user c, an object-based nearest neighbor algorithm considers those ratings $\{r(s_{\text{sim}},c)\}$ of items that are rated in similar way as item s, weighted by their similarity, when predicting $r(s,c)$. This approach is based on the assumption, that similar items are rated in a similar way by the users. [63] and [24] evidenced empirically that object-based algorithms are faster in computation with the same or even better quality of predictions than the best user-based approaches. Due to the same scalability issue as user-based approaches, [63] proposes to use only items co-rated by more than k users.

Probabilistic Approaches

In contrast to the memory-based approaches from above (using user-user similarities), model-based methods "use the collection of ratings to learn a *model*, which is then used to make rating predictions" [2]. Object-based nearest neighbor algorithms are an example of model-based CF methods, because (implicitly) a model of a user's rating behavior is created, taking into account the rating behavior of other users. There are two possible models proposed by [13] to determine this prediction: Bayesian networks and cluster models. The probability "User c will rate item s with r: $p(r|n,i)$ " is estimated and prediction is based on the most likely item or the predictand.

Problems

New items (as well as an insufficient number of items) are problematic for Collaborative Filtering systems, because if there are no ratings for a new item, it can't be recommended. A new user is also problematic for the system because there is no information about this user's preferences. There are several ways to handle this challenge of sparsity (of rating matrices), e.g. default voting, the use of user profile information (e.g. demographic similarity sometimes called "demographic filtering" [59]) or a dimensionality reduction technique (e.g. Singular Value Decomposition, SVD as used in [8, 64]).

1.2.3 Other Approaches

There are many other overlapping approaches and sub-approaches to Recommendation that emphasize certain flavors, paradigms, styles or techniques for constructing sub-systems of recommender systems.

Hybrid approaches try to combine the advantages of e.g. content-based filtering and collaborative filtering approaches to overcome their disadvantages [20]. Combined approaches use several recommenders in parallel, either mixing the the results and possibly additionally weighting them (e.g. by the recommendation multiplicity) or using the systems cascadingly [19].

Knowledge based recommender systems is a very general term for recommenders that use a formalization of domain knowledge to make or improve recommendations which is effectively true for most systems discussed here. However, the term may be justified if the focus is on exploiting explicitly encoded symbolic (e.g. ontological) knowledge (see e.g. [28]).

One example for knowledge based recommender systems are constraint based recommenders [28] where a formalization of the domain or more precisely the recommendation problem in form of constraints is constructed and constraint solvers are used to generate sets of allowable recommendations (that are compliant with the constraints).

Another variant are Rule-Based Recommender systems (see e.g. [1]), where a set of rules determines for a given user and a given context (previously seen items, current location, rating history etc.) what items to recommend. More specifically, rule-systems are an approach to implement the actual recommendation generation, once a meaningful rule-processable context-model has been computed (e.g. by constructing or assembling a user-item-rating-matrix, an item-item similarity measure, list of previously seen items etc. that can be used to formulate logical expressions for rules).

A third variant, Case-based Recommenders, focus on Case-Based-Reasoning, where a database of problems and corresponding solutions (recommendations) is kept and a new problem is compared via an appropriate similarity measure [23] to problems in the database and the best matching solution (recommendation) or a suitably transformed version of this solution (recommendation) is suggested.

Context-based recommender systems [3] are another important variation, where context is taken into account for improving recommendations. Context may be defined as in most cases highly dynamic, and implicitly gained data (plus a suitable semantics / model) that, if taken into account, may improve usefulness of services. Such services are then called context-aware [25]. (The concept of personalization, in contrast to that, usually involves additional data of lower dynamic nature and also explicitly stated data (mostly in form of a profile) in order to improve service usefulness). As an example for context-awareness, recommenders for points of interest may profit from the current location as context element. Besides individual context (location, disabilities etc.), social context may be of interest here too. Social context may encompass knowledge of social relations and their instantiations relevant for the current situation / service. Social recommenders can be viewed as context-aware recommender systems with an emphasis on social context. Here, the distinction between personalization and context is dropped and contexts on all dynamic scales (from long term social relations to short term social situations [34]) are considered. As an example consider a mobile restaurant recommender taking into account the presence of a vegetarian in the current social situation of a user. Besides taking social contexts into account, the term social recommender system may also be interpreted as being targeted towards social entities as recommended items (groups, experts, friends etc.). We will now discuss related work on both views of social recommender systems in more detail.

1.3 Social Recommender Systems

As has been said, in social recommender systems, aspects and components of traditional recommenders are explicitly designed using social entities and social contexts. One important variant of social recommender system (social filtering) is gained by substituting the user-neighborhood, whose ratings are considered to be similar to the current user's tastes or needs, by a model of social context [33]. Classic collaborative filtering (CF) may be already perceived as a social recommendation approach, since a set of users implicitly "works together" to produce useful recommendations. However, the social context exploited in classic CF is restricted to "similar rating behavior" which is a rather implict social context. In our Social

Recommendation approach discussed in sections 1.4 and 1.5, we aim at incorporating more explicit social contexts for recommendations.

The most basic easy distinction that can be made is based on the object of recommendation (*what* is recommended) and the method (*how* it is recommended). Both aspects may incorporate social context. We may distinguish item-like recommendation objects (texts, web-pages, films etc.) from social recommendation objects (persons, sub-networks of persons, groups of persons etc.). We may also distinguish item-based methods (user-item-matrix based collaborative filtering using on ratings of items, recommenders using knowledge about items, content-based recommenders using similarity between items etc.) from social methods (using sub-networks as neighborhoods, considering explicit user to user social relations etc.).

Examples can be:

1. Item Recommendation
   a. Classic collaborative filtering *(what: item-based, how: implicitly social)*
   b. Extended collaborative filtering by replacing the similarity-based neighborhood with a social neighborhood (social network) *(what: item-based, how: social)*
2. Person(s) Recommendation
   a. based on Social Network (e.g. Friend-Of-Friend Recommendation as in Facebook) *(what: social, how: social)*
   b. based on similar interest and tastes *(what: social, how: item-based)*
   c. Team Recommendation *(what: social, how: item-based or social)*

1.3.1 What: Item-Based, How: Social

In daily routine it is a common practice to ask friends and family for advice. This approach is easy to understand as trust in the source of a recommendation is essential for trust in the recommendation itself. As a user has more trust in recommendations from this social context he might also be willing to adopt recommendations of items outside his previous taste. Additionally a user's friends and family are likely

1. to know him and his tastes better than any "strangers" and
2. to have a similar taste and interests.

This brings up the assumption that they might make better / other or new types of recommendations, which we will elaborate on more thoroughly in later sections. Another aspect is that there is a strong correlation between friendship and geographic location. [45] found out that the probability of friendship in a social network increases with geographic proximity. In some domains geographic proximity can increase the quality of recommendations. Considering for example restaurant recommendations, people geographically near a user might probably make better recommendations as they know the restaurants and their relative quality in the corresponding area better than people thousand of kilometers away. Recommendations from friends might thus in many cases combine incorporation of local competence and social competence.

[66] picked up this approach and they were able to prove that a user's friends provide better recommendations than online recommender systems. The authors claim that the latter classic approaches are better in the field of making new, unexpected recommendations though. However, we argue analyzing our studies, that the new recommendations in social recommenders may be more likely to be accepted by the users, because the user is able to evaluate the implicit social context of a recommendation and it might be more interesting to investigate these unexpected recommendations in order to keep up with the current trends in his (peer-)group. We thus call them true horizon broadening recommendations and will discuss this effect late in more detail.

[11] also performed research on whether social context is important for recommendations. They were be able to show that "in taste domains, people prefer recommendations from people they know" [11].

[62] examined whether an underlying social network could improve recommendation quality. For this purpose they investigated the data of the movie recommender Filmtipset [29] which provides an underlying network. Their conclusion was, that "by utilizing user-user relations in a recommendation scenario one could improve the quality of recommendations".

In [68] a recommendation system for items (e.g. messages or documents) in a mobile personal information management system is implemented based on the current individual context (time and location and a user's personal ontology) as well as the current social context (the "socializable" parts of information spaces of available other users).

Social Filtering is also a good approach to overcome the problem of sparseness in collaborative filtering. If a user doesn't have enough co-rated items with enough users in the system, a trust based graph (a special case of social network) can be created and used to augment the number of similar users [51]. Social Recommenders are also a possibility to cope with the problem of *portfolio effect* (see Section 1.4, [32]).

These results beside others proved that it is worth to have a closer look at social collaborative filtering approaches. In sections 1.4 and 1.5 we will discuss these systems in more detail.

1.3.2 What: Social, How: Social

The probably best known recommender using social network to recommend other "social entities" (e.g. friends) is Facebook [26]. The main recommender approach of Facebook is the "Friend-of-Friend" (FoF) algorithm.

In general the FoF algorithm computes all friends of someone's direct friends and recommends those not already in his friend list. The more mutual friends this "friend-of-a-friend" and he have in common the higher ranked this "friend-of-a-friend" is. The assumption behind this approach is that a user might probably like those people who are friends of his friends. LinkedIn [48] e.g. has refined the social relations and uses "Colleague", "Classmate", "We've done business together", "Friend", "Other"," I don't know [name]" in order to make more precise recommendations.

Facebook developed several extensions to the FoF recommender, e.g. based on the ability to explicitly state participation at events. Another feature of Facebook is the mobile application called *places*. The intention is not to receive recommendations for previously unknown persons but to get recommendations of co-locations with known friends so a user is able to interact with them.

1.3.3 What: Social, How: Item-Based

Most of the known techniques to recommend items with the help of a content-based filtering approach can also be used to recommend users (possibly with some modifications). An interest-based person recommender may compare the profile (or other sources of interest and tastes) of a user c with all other users to determine a set of similar users. The most similar users are then recommended to user c.

Comparison of Social Network Based and Interest Based Algorithms. [21] compared in their study several approaches of social network based and interest based (content based) algorithms. They implemented an own people recommender system for Beehive (social networking site within IBM) using four different algorithms: Content Matching, Content-plus-Link (CplusL), Friend-of-Friend (FoF) and SONAR. Their recommender could be seen as a "social matching system" as described in [67]. Content-plus-Link is a content based algorithm enhanced by a social network, i.e. similar users are ranked higher if they have any social link. Their SONAR algorithm is based on the SONAR system "which aggregates social relationship information from different public data sources within IBM" [21], i.e. a relationship between two users can be detected when they interacted in the past (e.g. co-authoring, commenting ...). Each of these relationships between two users are then scored and all of them are aggregated to one single value. The user will be presented other users ordered by their final score.

In their analysis, [21] recommended 12 users (3 from each algorithm) to survey participants. Each recommendation contained a photo, job title, work location and an explanation why this person is recommended. The participants were asked to fill out a questionnaire afterwards [21]. The evaluation showed that pure content matching between users is the best in recommending new friends whereas the social network based approaches find more known people. The survey depicted also, that known recommendations are mostly rated well, which brings up the assumption that "the more known recommendations an algorithm produces, the more likely users are to consider those recommendations as good", whether "the more stranger an algorithms recommends, the more likely users will reject or not like the recommendations" [21], thus a combination of both ideas might be a good idea.

The evaluation of the survey also showed that there is an extensive need for explanations, why a person is recommended. Explanations are also beneficial for other types of recommenders as well and an interesting research topic in this field [53].

Expert Finders. Another field of recommending people based on content-similarity are Expert finders. Lin et al. proposed an expert finder with an integrated social networking system (SmallBlue). This framework combines the business intelligence of expertise (who knows what?) with a social network (who knows whom?), resulting in the ability for finding experts via datamining, information retrieval and social network analysis techniques [47], [46]. SmallBlue gathers its information amongst others from private sources as emails and chats, which has the advantage that it could be collected from every user and all information is automatically updated (e.g. through new emails). A big problem of using private data is that it may violate privacy, which is counteracted by SmallBlue with an innovative privacy approach. It relies only "on aggregated and inferred" private information [46]. If the privacy issue is not concerned, it would lead to rejection of the system by the users. SmallBlue uses also other (public) information sources as blog or forum posts inside the (company) network or articles written by "experts". With all the information taken into account, an expert for the searched topic could be found and recommended in most cases.

A big advantage of SmallBlue is the underlying network which presents also paths to reach the expert through own contacts based on the assumption that "In some companies it is a recognized and accepted practice to call other people to get information that is necessary for a project or assignment". [46]

1.3.4 Team Recommendation

Team Recommendation [14, 44, 50] goes further than expert finders described above. Of course one could implement a team recommender on the base of an expert finder via gathering single recommendations for all required positions in the team, but in this way one might not get the best possible team. One has to consider maximizing synergy effects in a team [16, 17, 15] and therefore will have to consider aspects like e.g. demographic variables (gender, age, location ...) or social influences [18]. We will introduce a few examples.

One point that needs to be considered in team recommendation is *motivation* of the team members. If they are motivated for a project, its output/outcome will be better. Of course one can motivate team members through a reward system, but it is also possible and even necessary to enhance motivation of team members by considering their *experience* to avoid demotivation by being underchallenged. On the other side, if a team member's competence is far behind his team colleagues, he might be overstrained and therefore also be demotivated. It is important to differentiate between knowledge and experience. If all team members have the same knowledge, information is redundant and there will be no innovation. If they are on a similar experience level but do not have the same knowledge, they have a common base to talk about and can thus learn from each other and improve their own knowledge and are for that reason potentially more motivated.

Trust is another variable important to consider at team recommendation. If team members trust each other, performance of the team will be better as if they don't count on each other. Absence of trust may lead to redundant procedures (e.g. "I am not sure if you have done this right, I better do it myself again."). Thus trust has a wide influence on effectiveness of a team.

[50] picked up the trust approach and extended the model of [27] "by incorporating trust into the recommender-based approach in order to add relational information" [50]. They conformed to [61] assuming that a single value can be used to express trust. A trust graph is a weighted directed graph. Even if user *c* has trust in user *d* it is not necessary that user *d* has also trust in *c*. The weights of the edges express trust. A trust based network can be compared with a social network. Instead of predicting if a user *c* will "like" user *d* (based on mutual friends, mutual interest ...) one may compute user *c*'s trust in *d* (possibly in relation to the trust *d* has got from other users). [50] differentiated between "direct trust propagation", "collaborative trust prediction" and "similarity-based trust prediction".

1.3.5 Recommendations for Groups

Another field of Social Recommenders are recommendations for Groups, which is covered in a special contribution in this book. One not only has to take into account a single user's preferences but those of a whole group which makes the recommendation process much more difficult. Making recommendations for groups one has to consider e.g. fairness. All members should be treated equally when making a recommendation. If user *c* of a group is not as satisfied with a recommendation as the rest of the group, one should possibly weight his opinion or requirements for the next recommendation in a higher way. Another aspect that arises when making recommendations for groups is that a user's preferences change according to the presence of other group members. A user might like horror films while being out with his friends whereas he might prefer comedy when going to cinema with his family. An example for a recommender that takes into account the change of preferences is FIT (Family Interactive TV), a TV program recommender outlined by [30].

One of the dimensions of group recommendation listed by [10] is "Number of recommendations per group". Treating group recommendations as sequences of recommendation "raises the issue of order and fairness". The order of recommendations is important because the first recommendation will influence the group members' preferences for the next recommendation. Having a sequence of recommendations one can treat each recommendation separately but then it is impossible to account for fairness as in the sense of above. Finally not knowing the length of sequence makes it necessary to regard every recommendation as a single problem but with respect to past recommendations. Another dimension that [10] listed is the "type of group". One needs to differentiate if a group is ephemeral or persistent and if it is public or private (with respect to privacy concerns) [55].

Before designing the generation of recommendations one should have a look at decision process in groups. For this we suggest to study the field of *"social choice theory"*. From this area three main voting strategies originate, which are applicable for group recommendations [52], which are [10]:

- *Average:* the average of the individual ratings for every item is computed and used as the item's ranking
- *Average Without Misery:* same as Average, but dismisses items with at least one individual rating below a certain threshold
- *Least Misery:* rank the items by their minimum individual ranking which takes into account, that a group is always as happy as the most unsatisfied group member

For computing group recommendations, [10] follow the classification of [39]:

- Merging of Recommendations Made for Individuals
- Aggregation of Ratings for Individuals
- Construction of a Group Preference Model / Aggregation of Profiles

For further research in the field of group recommendation we refer to [69, 38, 41, 40, 39, 52, 55, 22, 58].

1.4 Collaborative Filtering vs. Social Filtering: Taste Related Domains

Having laid out the ground by reviewing general approaches for recommender systems and having discussed related work on social recommender systems in more detail, we will now more thoroughly investigate Social Filtering. The present section will present a study on Social Filtering in taste related domains. In Section 1.5 we will then contrast this with another story, investigating the use of Social Filtering in a more factual domain. We will then compare both approaches.

Our first study [32] was based on two research questions:

1. Is rating behavior statistically dependent on friendship?
2. How does a social filtering approach perform compared to a collaborative filtering approach?

Basic Data Set Properties Our tests are build up on the data set of a German community Lokalisten [49], which is a Munich-based German language virtual community, focused on communication and spare-time. Lokalisten was founded in May 2005 and had (April 2007) approximately 700000 users all over Germany with a large and active local part based in Munich [49]. The community has a simple social network model where friendship relations are based on a two-way-handshake. Because Lokalisten has this strong user base in Munich we assumed that the users possibly will know the clubs in Munich and chose them hence as a possible domain of interest. With a breadth first search starting from one user we traversed the friendship graph until depth 4 resulting in 4249 users. After downloading the publicly available information we extracted nickname and friendship relations from the data set. We constructed an online-survey where 82 clubs in Munich could be rated on a discrete scale from 0 to 10. 0 was preselected and means that a user doesn't know the corresponding club. 1 was the worst possible rating and 10 the best. From all contacted user 1012 users (aged between 16 and 47) completed the online-questionnaire, resulting in 82 * 1012 = 82948 ratings.

Basic Notations and Approaches The retrieved (undirected) social relationship graph $G(U,E)$ (U : users, $E \subseteq \binom{U}{2}$: friendship relation) was stored in an adjacency matrix A_{ij}, where $A_{ij} = 1$ if $\{u_i, u_j\} \in E$, $A_{ij} = 0$ if $\{u_i, u_j\} \notin E$. The rating Matrix M_{ur} (82×1012) can be build with the 82-dimensional rating vectors of all users u_i as columns:

$$r^{(u_i)} = (r_0^{(u_i)}, r_1^{(u_i)}, r_2^{(u_i)}, \ldots r_{81}^{(u_i)}); r_k^{(u_i)} \in [0,10] \tag{1.1}$$

There are two different ways to build the user-user-similarity matrix $S_{ij} = sim(u_i, u_j)$ based on this ratings:

$$sim(u_i, u_j) = cos(r^{(u_i)}, r^{(u_j)}) = \frac{r^{(u_i)} \cdot r^{(u_j)}}{||r^{(u_i)}||||r^{(u_j)}||} \tag{1.2}$$

or

$$sim'(u_i, u_j) = sim(u_i, u_j)\, w_{co}(r^{(u_i)}, r^{(u_j)}) \tag{1.3}$$

The advantages of the cosine correlation are easier computation and no inadequately inclusion of "missing data" (i.e. zero values in M_{ur}). According to [57] both approaches should be equivalent. To take into account that similar vectors which contain e.g. only one single rating (e.g. $r^{(u_i)} = r^{(u_j)} = (0,0,\ldots,0,x,0,\ldots,0)$) would result in $sim(u_i, u_j) = 1$ while two vectors which contain a lot of (but not all) similar ratings would result in $sim(u_i, u_j) < 1$, the co-occurrence weight $w_{co}(r^{(u_i)}, r^{(u_j)})$ was introduced. The co-occurrence weight ensures that similarity's trustworthiness increases with the number of times the same item is rated by both users.

Experiment One

Our first experiment examined the statistical dependence of the rating behavior of the users and their social relations (groups and pairs). We assumed the use of cliques in the relationship graph for modeling social groups and extracted therefore pairs of friends and cliques for the group sizes of 3 to 6. To analyze if the ratings of friends are more similar than ratings of arbitrary groups (independent groups) we determined with a χ^2 -Test [12] if the independent variable "rating similarity" is statistically dependent of the variable "social relationships".

For the comparison of rating behavior of pairs of friends vs. non-friends-pairs, our H_0 Hypothesis was *"The rating-similarity is independent from the social relationship between rater-pairs"*. The result of the χ^2-Test was that H_0 has to be rejected, thus "rating similarity" and "social relationships"are statistically dependent.

We repeated the χ^2-Test for groups of users instead of pairs, to check whether the two variables are still dependent. In our data set the number of cliques larger than 4 was too small and therefore statistically not significant. We limited our test to examine the similarity of clique sizes 3 and 4 and independent sets of the same size.

For sets of user we had to define an average rating similarity: σ_U for a group U as

$$\sigma_U = \frac{2}{|U| * |U-1|} \sum_{i=0}^{|U-1|} \sum_{k=i+1}^{|U-1|} sim(u_i, u_k) \tag{1.4}$$

Our second H_0 Hypothesis was: *"The rating-similarity is independent from the social relationship between rater-groups"* . The result of the χ^2 -Test was again that H_0 has to be rejected, thus, for rater groups, the two variables are also statistically dependent. In our experiment we had also been able to prove that the correlation of 4-cliques is stronger than for 3-cliques which is in turn stronger than the correlation of pairs of friends. We found out that number of large cliques tend to be small, but we have clearly seen that groups of friends are more similar in their taste than arbitrary groups of people that do not know each other. This corresponds to observations made by [11] and [70]. In a taste related domain with a lower 'social' relevance the results may be different, but for example tastes in clothing, TV, cinema and music are strongly influenced by friends. Groups being "centers of taste" is a phenomenon which has been reviewed in [31] and is well known in social sciences. Among other reasons this is due to a "normative" effect that group-taste may have on members of the group [31].

As we pointed out in [31], several conclusions may be drawn from these empirical results:

- 'Virtual friend-relationships are capable of providing similar ratings in taste related domains
- Even binary friend-relations on average show more rating similarity than disconnected pairs
- Cliques tend to be centers of taste showing a higher similarity than mere friend-pairs on the average
- Based on similarity comparisons, cliques and friend-pairs might prove as suitable recommendation source since they share a common taste regarding the investigated domain.'

Experiment Two

In our second experiment we compared performance of collaborative filtering and social filtering. A classic collaborative approach consists of three basic steps. We have to compute:

1. similarities (matching)
2. correlation-thresholding (neighborhood creation)
3. weighted average ratings (prediction computing)

To measure the quality of the generated predictions (here: club-ratings) we compared them with the true values (the true ratings). The same three steps are, with a differing step 2 (neighborhood creation), applicable to a simple social filtering approach. For the neighborhood creation we used the retrieved social relations instead of correlation-thresholding. As already mentioned, we have gained a rating vector from each user u_i: $\mathrm{r}^{u_i} = (r_0^{(u_i)}, r_1^{(u_i)}, ..., r_{81}^{(u_1)})$ $(r_k^{(u_1)} \in [0, 10])$. A recommender can (in principle) predict any of these ratings denoted as $\mathrm{pr}^{u_i} = (pr_0^{(u_i)}, pr_1^{(u_i)}, ..., pr_k^{(u_1)})$ with $0 \leq k \leq 81$. A prediction can be *adequate, inadequate* or *novel*. On the basis of these three sets we defined *precision, recall, f-measure* and *mean absolute error (MAE)* in a standard way to have measures for comparing. As explained above, the neighborhood creation of conventional CF systems is threshold based. That is, to compute the predictions (recommendations) for a user u_i one considers all users u_j with a rating similarity $S_{ij} = sim(u_i, u_j) \geq \lambda$. The neighborhood $N_{coll}^{u_i}$ for a user u_i can then be defined as $N_{coll}^{u_i} = \{u_j | S_{ij} \geq \lambda\}$, see [31]. While we used the cosine-based correlation in the first experiment (see eq. (1.2)), the second experiment required the use of the co-occurrence weighted similarity (see eq. (1.3)) to optimize the CF algorithm [37].

Since the threshold-based generated neighborhoods can (depending on λ) be much larger than the set of friends of a user, the neighborhood for social filtering was extended to the set of friends *and* friends-of-friends of a user.

$$N_{\text{social}}^{(u_i)} = F^{(u_i)} \cup \{u_k \mid \exists u_j \in F^{(u_i)} : A_{jk} = 1\}. \tag{1.5}$$

With those neighborhoods prediction-vector for a user u_i can be defined in three ways:

1. simple averaging of ratings

$$r^{(u_i)} = \frac{1}{|N^{(u_i)}|} \sum_{\{j|u_j \in N^{(u_i)}\}} r^{(u_j)} \tag{1.6}$$

2. similarity weighted average over the rating-vectors *(to overcome the different rating bias)*

$$pr_m^{(u_i)} = \overline{r^{(u_i)}} + \frac{1}{\sum_{\{j|u_j \in N^{(u_i)}\}} S_{ij}} \sum_{\{j|u_j \in N^{(u_i)}\}} (r_m^{(u_j)} - \overline{r^{(u_j)}})\, S_{ij} \tag{1.7}$$

3. no similarity weighting *(necessary for a fair comparison as the weighted rating similarity for N_{social} needs to be replaced by the strengths of a social relation, which was in our case not available)*

$$pr_m'^{(u_i)} = \overline{r^{(u_i)}} + \frac{1}{|N^{(u_i)}|} \sum_{\{j|u_j \in N^{(u_i)}\}} (r_m^{(u_j)} - \overline{r^{(u_j)}}) \tag{1.8}$$

$\overline{r^{(u)}}$ is the average rating for a user u.

To be able to determine whether the computed predictions are adequate, inadequate or novel, we created 25 "sparse" version of the rating matrix M_{ur}, by randomly choosing $n * 1000$ ($n \in \{1, 2, \ldots, 25\}$) ratings in M_{ur}.

Considering the f-measure results some interesting aspects has shown up:

- For *high sparseness* (few ratings), *simple averaging prediction calculation* (eq. (1.6)) is best by far
- For *low sparseness* (many ratings), simple averaging is beaten by *social neighborhood with non-similarity weighted* (eq. (1.8)) and *similarity weighted* (eq. (1.7)) predictions computing

Similarity weighted collaborative filtering is one of the best neighborhood based filtering algorithms [37] so the social approach works excellently.

As we analyzed in [31], while the peer-group of friends and friend-friends *as a whole* can be a good indicator of personal taste even in very sparse situations, *single* friends and friend-friends may have a different taste then the active user which becomes especially apparent in the highly sparse situations where few ratings are available. For the collaborative approach we found out that for low sparseness the similarity weighted (eq. (1.7)) and the non-similarity weighted (eq. (1.8) approaches perform only slightly better than the *worst* social version (simple averaging, (eq. (1.6)), which is in turn useless for the collaborative approach (also showed by [37]). A disadvantage of collaborative approach is that it is very sensitive to the λ-threshold. It is possible to fine tune λ by observing the system but as optimal λ value increases substantially with decreasing sparseness it might be too hard to determine in a real recommender system. This also substantially supports the social approach.

The MAE results showed us very similar results. Only the collaborative approaches similarity weighted (eq. (1.7)) and non-similarity weighted (eq. (1.8)) have been able to produce as accurate prediction as the social approach (for most levels of sparseness). But the price for accuracy is less novelty. For high sparseness (i.e. a very low λ resp. a large neighborhood) there a many (but very inaccurate) novel predictions, but for lower sparseness (i.e. a higher λ resp. a smaller neighborhood) the number of novel prediction has been already 25% lower than for social neighborhoods. Social approaches started with a few number of novel ratings at high sparsity but increased this number substantially. This is a clear advantage of the social approach.

As already mentioned in Section 1.3.3, [21] showed that there is an extensive need for an explanation why someone or something is recommended. "Collaborative systems today are black boxes, computerized oracles which give advice but cannot be questioned" [54]. Social recommenders may overcome this lack of transparency and give an explanation for the recommendation because its origin (the set of users used) is known. Thus a higher level of trust is reached for Social Recommenders.

1.5 Collaborative Filtering vs. Social Filtering for Discussion Boards

Having seen that Social Filtering is a successful approach in domains that are related to personal taste, we will now investigate the suitability of social recommender systems (and especially Social Filtering) for discussion boards with an integrated social network. User data and discussion board data of an existing network with half a million of users were recorded and a survey was carried out amongst a part of the users to gather explicit ratings. Then different collaborative filtering and social filtering approaches were implemented and compared in two different test procedures based on the network data. The traditional collaborative filtering approaches show higher performances in both test procedures than the social filtering approach. Thus the hypothesis that traditional approaches can be optimized through the use of the social network could not be confirmed. We discuss the results of the test procedures and compare them with results of the preceding study and point out possible explanations for the weak performance of the social approaches.

User-based Nearest Neighbor Algorithm For our study we used *user-based nearest neighbor* algorithm (henceforth denoted as *NN algorithm*) and will describe it now in detail. The decision for this algorithm is partly based on the results of [13] and partly on the limited scope of this study which made it impossible to implement all potential algorithm candidates. Nevertheless the user-based nearest neighbor algorithm is without a doubt considered to be one of the most typical representatives of collaborative filtering algorithms.

As has partly been introduced in previous sections, the NN algorithm tries to copy the real life behavior we already just mentioned when introducing collaborative filtering by defining a neighborhood $neighbors_n$. If the algorithm wants to predict the utility of an item i for an user n it uses the ratings of the users included in $neighbors_n$ to generate a prediction $pred_{n,i}$. One variable of the algorithm is the way how $neighbors_n$ is defined. Most simply it is possible to define *all* users but user n as neighbors of n, but it is more common to define a *similarity function* $sim_{n,u}$ that denotes the similarity between two users n and u, so that $neighbors_n$ could be defined as

$$neighbors_n = \{u \mid u \in N \wedge sim_{n,u} > \lambda\} \tag{1.9}$$

where λ is a chosen threshold value.

Often (and also in the implementations in this part of our contribution) the similarity function is defined as Pearson correlation (according to [60]) by comparing all items $CR_{n,u}$ (denoted as C in the equation below due to limited space) that both n and u have rated already:

$$sim_{n,u} = \frac{\sum_{i \in C}(r_i^{(n)} - \overline{r^{(n)}})(r_i^{(u)} - \overline{r^{(u)}})}{\sqrt{\sum_{i \in C}(r_i^{(n)} - \overline{r^{(n)}})^2}\sqrt{\sum_{i \in C}(r_i^{(u)} - \overline{r^{(u)}})^2}} \tag{1.10}$$

A second variable of the algorithm is the way the the neighbor's ratings are used to predict the utility of item i for user n. We extend the naive approach of simply averaging over all ratings of the neighbors,

$$pred_{n,i} = \frac{\sum_{u \in neighbors_n} \cdot r_i^{(u)}}{|neighbors_n|}, \tag{1.11}$$

in a standard way (also referenced before) by considering average ratings (some users may give high ratings more often than others), [65]

$$pred_{n,i} = \overline{r^{(n)}} + \frac{\sum_{u \in neighbors_n}(r_i^{(u)} - \overline{r^{(u)}})}{|neighbors_n|}, \tag{1.12}$$

and considering the different weights of similarity (not all users are equally similar to user n):

$$pred_{n,i} = \overline{r^{(n)}} + \frac{\sum_{u \in neighbors_n} sim_{n,u} \cdot (r_i^{(u)} - \overline{r^{(u)}})}{\sum_{u \in neighbors_n} sim_{n,u}} \tag{1.13}$$

Social Filtering As introduced above, Social Filtering is associated with the integration of an underlying social network into the calculation of the predictions of a recommender system. This is most commonly done by simply changing the definition of $neighbors(n)$ so that the social ties play a prominent role in the neighborhood of n. For example $neighbors_n$ could be defined as all users who have a direct friendship relation to user n. But there are also variations possible, for example to keep the original definition of $neighbors_n$ but restrict the possibly selected users to those who are within the social network of user n.

As shown above, users trust recommendations made by their friends more than recommendations made by an online recommender system.

Analogous to the study presented in previous sections, in this study we assume that recommendations made by an online recommender system for discussion boards can be improved by using an underlying social network and social neighborhood. Therefore we conducted an empirical study which we will present in the following.

1.5.1 Empirical Study

Data Set Properties

A possible data set should fulfill three requirements:

1. a set of users who are interconnected to each other by a social network in order to be able to judge the influence of social structures on the users' ratings
2. threads/topics which contain user written posts
3. ratings of threads by users

An artificial data set was ruled out for this would have modeled reality insufficiently and there is plenty of appropriate data to be found on the Internet. We chose a German Internet community which had a community wide discussion board. Moreover the community had no major privacy settings so that user profiles, friendships relations and the complete content of the discussion board could be downloaded. This resulted in 518.211 user profiles, interconnected by 1.107.928 friendship relations and – with regard to the discussion board – 207.180 posts in 4.354 threads.

So far requirements 1 and 2 from the list above are fulfilled but ratings of threads by the users are still required. As already mentioned there are principally two ways to define ratings – implicit and explicit – and both rating variants were collected. We defined as implicit rating $r_{impl_i}^{(n)}$ of user n for thread i simply the number of posts that n has written in i.

By far most users had less than 100 posts in total so that (when comparing that number to the number of several thousands of threads) it becomes clear that this results in a pretty sparse rating environment. So additionally we tried to collect explicit ratings by inviting members of the community to participate in an online survey which allowed the participator to rate 50 discussion board threads on a scale from 0 ("I'm not interested at all") to 6 ("I find this very interesting"). These 50 threads were randomly chosen before (and were *the same for all* participators) so that we ended up with a subset of threads for which explicit ratings by a subset of the user base were available. Eventually a total of 501 users took the time for filling out our online survey.

Design of the Experiments

Basic Considerations The design of the experiments is strongly influenced by the suggestions for evaluating recommender system in [35]. Regarding the user problem we investigated the recommender system's ability to solve two different problems (which are named as in [35] where the scenarios are originally proposed):

- *Annotation in Context*
  All threads should be annotated with hints (for example five star ratings) for the user that indicate how interesting the user may find the thread. This comes down

to the question whether the recommender system is able to accurately predict the rating a user would give an item if he had to rate it explicitly.

- *Find Good Items*
In this scenario the most interesting threads out of all threads in the forum shall be presented to the user, e.g. in a list. This implies that the recommender system is able to find all relevant threads in a huge pile of mostly uninteresting items.

The previously collected explicit ratings seem well suited for an investigation of the first problem scenario because we have a set of threads for which we exactly know the explicit ratings at least for a subset of the users so that we can compare these actual ratings with the predictions made by the recommender system (*Experiment One*). The implicit ratings, i.e. the number of posts of a user in a thread, can serve as a basis for testing a recommender system with regard to the second problem scenario by examining whether the recommender system is able to identify exactly those threads in which the user has posted at least once if we assume that posting in a thread indicates relevance of this thread to the user (*Experiment Two*).

We now introduce some formal definitions to be able to discuss the experiment analysis below. Let N be the set of all users and N_U the set of users who participated in the online survey ($|N| = 518.212$ and $|N_U| = 501$). Furthermore, let T be the set of all threads ($|T| = 4.354$), let P be the set of all posts ($|P| = 207.180$) and let $P_t \subset P$ be the set of posts in $t \in T$. For a thread $p \in P$ we denote $p_{author} \in N$ as the author of the thread and $p_{thread} \in T$ as thread $t \in T$ for which $p \in P_t$ and $p_{rank} \in \mathbb{N}$ the position of the post in vector $\overrightarrow{p_{thread}} = \{p_1, p_2, \ldots, p_{|P_{p_{thread}}|}\}$ (where $\overrightarrow{p_{thread}}$ is the chronological order of posts in p_{thread}.

For every user $n \in N$ a vector

$$r^{(n)}_{impl} = (r^{(n)}_{impl_1}, r^{(n)}_{impl_2}, \ldots, r^{(n)}_{impl_{4354}}) \tag{1.14}$$

exists where $r^{(n)}_{impl_i} = |\{p \in P_i \mid p_{author} = n\}|$. In the online survey we randomly selected a set $T_U \subset T$ with $|T_U| = 50$. Therefore for every user $n \in N_U$ also a vector

$$r^{(n)}_{expl} = (r^{(n)}_{expl_1}, r^{(n)}_{expl_2}, \ldots, r^{(n)}_{expl_{50}}) \tag{1.15}$$

with $r^{(n)}_{expl_i} \in [0, 1, \ldots, 6]$ exists.

Basic Properties of All Tested Algorithms. Except for three naive algorithms that we will explain below, all tested algorithms are variations of the NN-algorithm. The variations are what makes the empirical study interesting by for example using the social network for the generation of $neighbors_n$ and comparing the results of this algorithm to the results of a similar algorithm which does the same but generates $neighbors_n$ on basis of a similarity function sim. If the algorithm that uses the social network would perform significantly better than the other algorithm (which does not use the social network) then this would be a strong argument for our hypothesis that the use of an underlying social network is able to improve the performance of recommender systems. This approach of altering single variables of the same algorithm allows for a much better discussion of the results because the total configuration of all variables strongly influences the performance of such a relatively complex algorithm.

So for the rest of subsection 1.5.1 we will point out the single variables of the NN-algorithm which we alter later in the experiments to end up with different variations.

Calculation of Prediction. Principally there are three different possible approaches to calculate the actual prediction $pred_{n,i}$ for an user n and an item i:

1. First approach: simply take the average of the appropriate ratings of users in $neighbors_n$ (cp. equation (1.11))
2. Second approach: first approach plus consideration of average ratings of users (cp. equation (1.12))
3. Third approach: second approach plus consideration of similarity values (cp. equation (1.13))

All three approaches require a definition of $neighbors_n$ and some require a similarity matrix.

Similarity Matrix. Those approaches that base their definition of $neighbors_n$ on a similarity function (cp. equation (1.9)) and in any case the third approach above require a similarity matrix

$$M_S = \begin{pmatrix} sim_{n_1,n_1} & \cdots & sim_{n_1,n_{|N|}} \\ \vdots & \ddots & \vdots \\ sim_{n_{|N|},n_1} & \cdots & sim_{n_{|N|},n_{|N|}} \end{pmatrix} \tag{1.16}$$

for the calculation of a prediction $pred_{n,i}$.

sim can be defined by the Pearson correlation (cp. equation (1.10), this is the case in *Experiment One*) or by other means (this is explained in detail when introducing *Experiment Two* in Section 1.5.1.2)

Calculation of Neighborhood. The set of neighbors $neighbors_n$ (which is used in the calculation of $pred_{n,i}$) can actually be defined on the basis of M_S but it is also possible to use the social network of a user. Because the definition of $neighbors_n$ on basis of the social network is the same in both experiments (in contrast to the definition on basis of M_S due to the different definitions of *sim* in the two experiments) a description of this definition will be given here:

Let $G(U,E)$ be the graph of a social network where U are the users (nodes) and $E \subseteq \binom{U}{2}$ are the friendship relations (edges). Assuming that the friendship relations are undirected (as they actually are in the data set we use), i.e. $(n,u) \in E \Leftrightarrow (u,n) \in E$, then an adjacency matrix exists:

$$M_A = \begin{pmatrix} A_{n_1,n_2} & \cdots & A_{n_1,n_{|N|}} \\ \vdots & \ddots & \vdots \\ A_{n_{|N|},n_1} & \cdots & A_{n_{|N|},n_{|N|}} \end{pmatrix} \tag{1.17}$$

$$A_{n,u} = \begin{cases} 1 & if\ (n,u) \in E \\ 0 & if\ (n,u) \notin E \\ 0 & if\ n = u \end{cases}$$

This notation allows us to define neighborhood on the first friendship level as

$$neighbors_n = \{u \mid N_{n,u} = 1\}. \tag{1.18}$$

Of course it is also possible to widen the circle of friendship and define neighborhood with two friendship levels as

$$\begin{aligned} neighbors_n = &\{u \mid A_{n,u} = 1\} \\ &\cup\{u \mid \exists v : A_{u,v} = 1 \wedge A_{n,v} = 1\}. \end{aligned} \tag{1.19}$$

A possibly interesting structure for defining a neighborhood comes from the field of social network analysis. A *clique* is a subset $U \subseteq V$ of a graph (V,E) that induces a subgraph $G(U)$ so that every node $v \in U$ is connected with every other node $u \in U$ (for a more detailed description of cliques, please have a look at [43]). There are different ways in which one could use this structure for the definition of $neighbors_n$. We decided for a definition where a user u belongs to $neighbors_n$ if he is in *any* kind of clique with user n, i.e. it is sufficient to check for 3-node-cliques:

$$\begin{aligned} neighbors_n = &\{u \mid A_{n,u} = 1 \\ &\wedge \exists v : A_{u,v} = 1 \wedge A_{n,v} = 1\} \end{aligned} \tag{1.20}$$

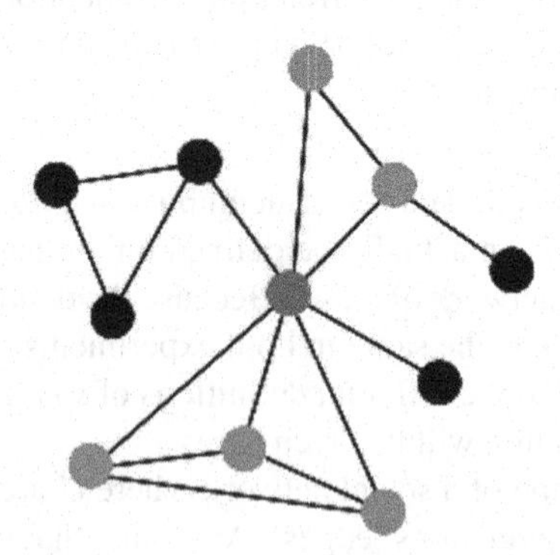

Fig. 1.1 Visualization of $neighbors_n$ when considering cliques

You can see an example for this definition of $neighbors_n$ in Figure 1.1. The red node marks user n within a social network. Only the green marked nodes would be part of $neighbors_n$ because only those are in a clique with the red node (the bottom nodes even form a 4-nodes-clique). The black marked nodes which have a direct connection to the red node plus the green marked nodes would form $neighbors_n$ if we defined $neighbors_n$ according to equation (1.18). Finally if we defined $neighbors_n$ according to equation (1.19) all nodes would be part of $neighbors_n$.

Naive Algorithms. To test the other algorithms against most naive predictions we implemented two naive algorithms. The first one, called *Random,* randomly selects a value on the possible rating scale $r \in [a_1, a_2]$ in order to generate a prediction $pred_{n,i}$ for user n and item i:

$$pred_{n,i} = random(a_1, a_2) \tag{1.21}$$

The second one, *default prediction*, works even simplier: instead of randomly selecting a value on the rating scale for every new prediction only one value is predefined and assigned to *all* predictions (we chose a default prediction value of 3):

$$pred_{n,i} = r = 3 \tag{1.22}$$

1.5.1.1 Experiment One

As already explained in 1.5.1 we splitted the data set in order to conduct two different test procedures. *Experiment One* uses only the explicit ratings and the threads in T_U (the 50 randomly selected threads in the online survey) and models the user problem *Annotation in Context*, whereas *Experiment Two* uses only implicit ratings and all threads to investigate the user problem *Find Good Items*.

So in *Experiment One* the goal was to find a good recommender algorithm that accurately predicts the explicit ratings which the participators in the online survey gave the 50 threads in T_U. Several different collaborative filtering algorithms plus two naive algorithms were implemented and tested to be able to compare common collaborative filtering approaches with social filtering approaches.

Most recommender algorithms need training data for building up an internal knowledge model that is used to predict unrated items for a particular user. Therefore T_U was divided randomly in two subsets T_{train} and T_{test} ($|T_{train}| = |T_{test}| = \frac{|T_U|}{2}$ and $T_{train} \cap T_{test} = \varnothing$). The ratings of threads in subset T_{train} served as training data for the algorithms which then had to predict the ratings for all threads in subset T_{train} in the second part of the test.

Those algorithm variants that need a similarity matrix M_S calculated it by means of the Pearson correlation (cp. equation (1.10)) and used only the threads in T_{train} for this calculation. Again those algorithm variants that base their definition of neighborhood on M_S defined $neighbors_n$ as in equation (1.9). This could disadvantage those algorithm variants if we chose λ poorly. Hence every run of such an algorithm was done 10 times with different values of $\lambda \in \{0, 0.1, \ldots, 0.9\}$ and only the results of the best run were fixed.

After the calculation of a neighborhood every algorithm eventually calculated a prediction for every user $n \in N_U$ and every thread $t \in T_{test}$. For this calculation only explicit ratings for threads in T_{test} were used. This resulted in a prediction matrix M_P for every algorithm:

$$M_P = \begin{pmatrix} pred_{n_1,t_1} & \cdots & pred_{n_1,t_{25}} \\ \vdots & \ddots & \vdots \\ pred_{n_{501},t_1} & \cdots & pred_{n_{501},t_{25}} \end{pmatrix} \tag{1.23}$$

$$n \in N_U, t \in T_{test} \ : \ pred_{n,t} \in [0,6]$$

We decided for the f-measure and the mean average error (MAE) as standard tools for measuring the performance of recommender algorithms (see [35] for a detailled description if you are not familiar with these measures). So the f-measure was defined as

$$F = \frac{2 \cdot Precision \cdot Recall}{Precision + Recall}, \tag{1.24}$$

where *Precision* and *Recall* were defined as follows:

$$\begin{aligned} T_{selected} &= \{(n,t)\ n \in N_U, t \in T_{test} | pred_{n,t} \geq \gamma\} \\ &\gamma = 4 \\ T_{relevant} &= \{(n,t)\ n \in N_U, t \in T_{test} | r^{(n)}_{expl_t} \geq \gamma\} \\ T_{accurate} &= \{(n,t) \in T_{selected} | r^{(n)}_{expl_t} \geq \gamma\} \\ T_{inaccurate} &= T_{selected} \setminus T_{accurate} \\ Precision &= \frac{|T_{accurate}|}{|T_{accurate}| + |T_{inaccurate}|} \\ Recall &= \frac{|T_{accurate}|}{|T_{relevant}|} \end{aligned}$$

MAE was defined as follows:

$$MAE = \frac{1}{|T_{test}| \cdot |N_U|} \cdot \sum_{n \in N_U} \sum_{t \in T_{test}} |r^{(n)}_{expl_t} - pred_{n,t}| \tag{1.25}$$

It is unrealistic for a scenario which consists of 50 items that every user has rated all 50 items. Therefore the rating vectors were sparsed out artificially, i.e. some ratings were deleted. Let $\alpha \in [1, 2, \ldots, 20]$ be the *sparsity factor*, then the rating vector $r^{(n)}_{expl}$ of every user is transformed into a sparsed out variant $\tilde{r}^{(n)}_{expl}$ which contains only $\frac{\alpha}{20}$ of the original ratings (selected on a random basis). Every algorithm was tested 20 times with a different sparsity factor each time (1 to 20).

Moreover, every run was conducted 10 times and an average was calculated over all results since chance plays a prominent role in this scenario with only 25 test items. By averaging over 10 runs the results could be smoothened effectively without affecting validity.

We now give a list of the implemented algorithms that were tested in *Experiment One*:

- **random**
  neighborhood: n/a
  similarity matrix: n/a
  calculation of prediction: cp. (1.21)
- **default prediction**
  neighborhood: n/a
  similarity matrix: n/a
  calculation of prediction: cp. (1.22)
- **standard collaborative**
  neighborhood: via similarity matrix (cp. (1.9))
  similarity matrix: via pearson correlation (cp. (1.10))
  calculation of prediction: consideration of average ratings and similarity weights (cp. (1.13))
- **standard collaborative w/o similarity**
  neighborhood: via similarity matrix (cp. (1.9))
  similarity matrix: via pearson correlation (cp. (1.10))
  calculation of prediction: consideration of average ratings (cp. (1.12))
- **standard collaborative simple averaging**
  neighborhood: via similarity matrix (cp. (1.9))
  similarity matrix: via pearson correlation (cp. (1.10))
  calculation of prediction: simple average (cp. (1.11))
- **social 1-level**
  neighborhood: 1st level of friendship (cp. (1.18))
  similarity matrix: via pearson correlation (cp. (1.10))
  calculation of prediction: consideration of average ratings and similarity weights (cp. (1.13))
- **social 2-level**
  neighborhood: 2nd level of friendship (cp. (1.19))
  similarity matrix: via pearson correlation (cp. (1.10))
  calculation of prediction: consideration of average ratings and similarity weights (cp. (1.13))

- **social 2-level w/o similarity**
  neighborhood: 2nd level of friendship (cp. (1.19))
  similarity matrix: n/a
  calculation of prediction: consideration of average ratings (cp. (1.12))
- **social simple clique**
  neighborhood: cliques (cp. (1.20))
  similarity matrix: via pearson correlation (cp. (1.10))
  calculation of prediction: consideration of average ratings and similarity weights (cp. (1.13))

The results of the different algorithms with regard to the perfomance measures are presented in Figure 1.2.

Discussion of Experiment One

The following discussion refers to Figure 1.2. First of all it is apparent that both naive algorithms, *random* and *default prediction* have constant curves. This is consequential because the sparsity factor α does not play any role at all for the predictions of both algorithms. The different constant values can be reconstructed easily with respect to the definitions of the algorithms and the performance measures (e.g. a default prediction of 3 obviously was closer to the real distribution of ratings as the average of randomly chosen ratings, resulting in a lower MAE value for *default prediction*).

Regarding the three standard collaborative filtering algorithms one can see that two of these, *standard collaborative* and *standard collaborative w/o similarity*, have very similar curve progressions both in the f-measure and the MAE and both perform best in both performance measures compared to all other tested algorithms. Without going into detail we point out that the eye-catching jump concerning f-measure from $\alpha = 0.05$ to $\alpha = 0.1$ is due to probability reasons caused by the test design. Though the third algorithm, *standard collaborative simple averaging*, performs clearly worse than the other two and it even cannot compete with the *random* algorithm with respect to the f-measure it does not perform significantly worse than the non-standard collaborative filtering approaches concerning the f-measure and performs even significantly better concerning the MAE. These first observations imply that

1. the use of similarity weights at the calculation of predictions seems to be rather unimportant for recommender performance (and gets even less important the lower sparsity is) while considering average ratings seem important.
2. two of the standard collaborative filtering algorithms clearly outperform all other algorithms (including the social filtering approaches). Hence in this test scenario social filtering approaches do not seem to be able to improve recommender performance.

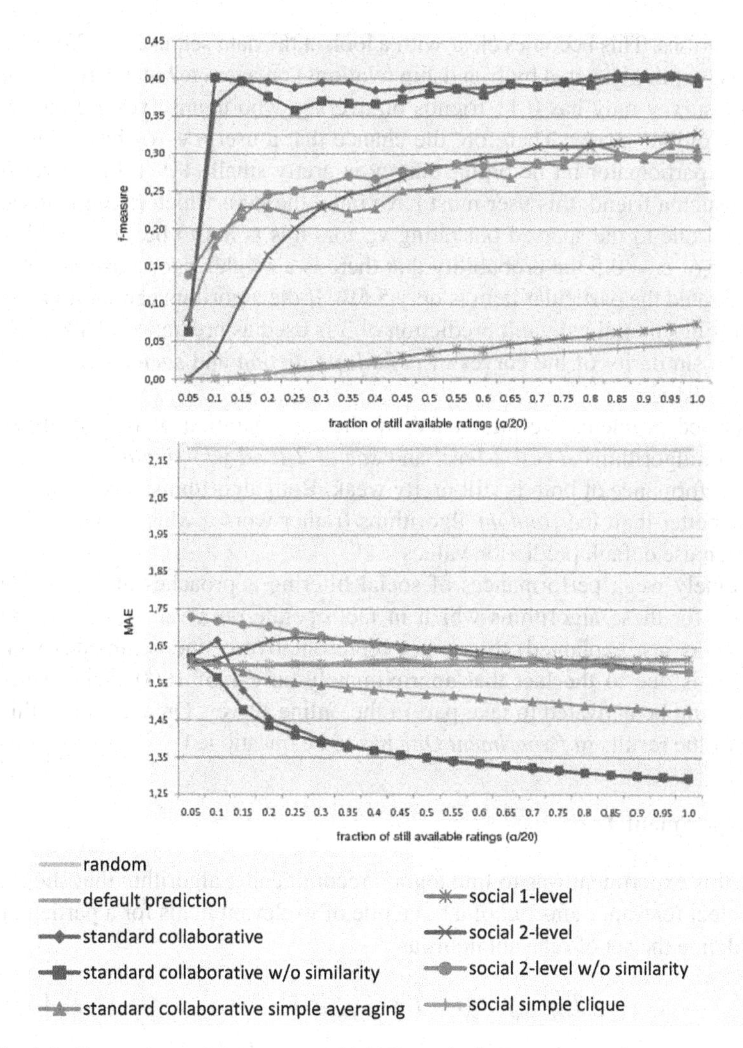

Fig. 1.2 Comparison of f-measure and MAE results in Experiment One

Additionally it is interesting that both best performing algorithms use the additional information that comes along with higher α for more precise predictions (hence the decreasing MAE values) but are not able increase the f-measure values. This indicates a weakness of the algorithms at the transformation of their predictions on the binary scale *select* or *do not select*. However, the main goal in this part of our contribution is to investigate the potential of social filtering approaches compared to traditional collaborative filtering approaches so that we will not go into further details with that problem.

The weak performance of *social 1-level* and *social simple clique* is striking. The reason for this weak performance lies in the fact that both operate on the first level of friendship relations. This becomes clear with a look at the data set: users in N_U have 17.07 friends (explicitly stated by friendship relations) on average but a participator in the online survey only has 0.11 friends on average who themselves are participators in the online survey. Therefore the chance that a user $n \in N_U$ has a friend who is also a participator int he online survey is pretty small (11%). Moreover *if* the user has such a friend, this user must have rated the item which rating is to be predicted (and due to the sparsed out rating vectors this is not to be sure). With a sparsity factor of $\alpha = 0.5$ the probability that there is a neighbor for a user $n \in N_U$ who has also rated the particular item is only 5.5%. If the algorithm cannot find any neighbors' ratings than the default prediction of 3 is used as prediction. This is the reason for the similarity of the curves of *default prediction* and *social 1-level* and *social simple clique*.

The described problems are attenuated by the consideration of two levels of friendship, i.e. algorithms *social 2-level* and *social 2-level w/o similarity*. Nevertheless the performance of both is still pretty weak. Both algorithms never perform significantly better than the *random* algorithms (rather worse) which implies that both also often use default prediction values.

The extremely weak performances of social filtering approaches may indicate disadvantages for these algorithms which in fact operate on a very sparse social network (as was just explained) that is not representative of the complete social network. This is due to the fact that approximately only 1 of 1000 users of the community could be activated to take part in the online survey. On account of this the validity of the results in *Experiment One* has to be questioned.

1.5.1.2 Experiment Two

The goal in this experiment was to find a good recommender algorithm that should be able to select relevant items out of a huge pile of irrelevant items for a particular user n. We define the set of relevant items as

$$T^{(n)}_{relevant} = \{t \in T \mid r^{(n)}_{impl_t} \geq 1\}. \tag{1.26}$$

In the test procedure every algorithm has to select z_n threads with the highest predictions out of a subset $T^{(n)}_{test} \subset T$ for every user n, whereas z_n denotes the number of relevant items for a user n (i.e. $r = |T^{(n)}_{relevant}|$) to allow the algorithms to find all relevant items. We denote $T^{(n)}_{selected}$ as the set of the threads the algorithm selects. $T^{(n)}_{test}$ consists of relevant and irrelevant items:

$$\begin{aligned}
T^{(n)}_{test} &= T^{(n)}_{relevant} + T^{(n)}_{false} \\
T^{(n)}_{false} &\subseteq T^{(n)}_{irrelevant} \\
T^{(n)}_{irrelevant} &= T \backslash T^{(n)}_{relevant} \\
|T^{(n)}_{false}| &= \nu \cdot |T^{(n)}_{relevant}|, \; \nu \in \mathbb{N}
\end{aligned} \tag{1.27}$$

The larger the value v is chosen the more difficult it gets for an algorithm to select the relevant items out of $T_{test}^{(n)}$. Every algorithm was tested with 9 different $v \in [1,\ldots,9]$. Except for *standard collaborative*, *social 2-level w/o similarity* and *social simple clique* every run was conducted 5 times and the results were averaged. For the three mentioned exceptions this could not be done due to time limits and the massive calculations needed for a single run alone (but this should not bother the validity as the influence of chance should be much smaller here compared to experiment one because much more than 25 items are used as test subset).

When designing the test procedures we encountered a problem that could bias the results: A separation of all items in two subsets (training and testing set) for those algorithms which need training data would not lead to meaningful results because of the high sparsity of implicit ratings (users are active in only a pretty small fraction of threads). A solution to the problem of high sparsity is to allow the algorithms to train with all implicit ratings available. But it is not desirable that if the algorithm calculates a prediction $pred_{n,t}$ it has already used the information that user n has posted in topic t for the generation of a similarity matrix. This would most definitely bias the results. Hence we defined that when calculating a prediction $pred_{n,t}$ for a *relevant* thread an algorithm had to use an altered rating vector $\tilde{r}_{imp}^{(n)}$ which is the same as the original rating vector but without the positive rating for thread t, i.e. $r_{impl_t}^{(n)} = 0$, for the calculation of similarities. On the one hand this increased the necessary calculation power dramatically, on the other hand it was only required for predictions for relevant items because for irrelevant items there is not even any information that could be deleted.

Due to the minimal differences in *Experiment One* only one variant of common collaborative filtering algorithms (the one we assumed to be the one with the highest performance) was implemented, i.e. *standard collaborative* from Experiment One. But instead of the pearson correlation as a means for defining similarity *standard collaborative* in *Experiment Two* uses another definition of similarity:

$$sim_{n,u} = |\{t \in T \mid r_{impl_t}^{(n)} \geq 1 \wedge r_{impl_t}^{(u)} \geq 1\}|, \tag{1.28}$$

i.e. the number of threads which both have actively participated in. The definition of neighborhood does without a threshold value as in equation (1.9):

$$neighbors_n = \{u \in N \mid sim_{n,u} > 0\} \tag{1.29}$$

Apart from these specific definitions *all* algorithms do not use the actual value of $r_{impl_t}^{(u)}$ for the calculation of a prediction $pred_{n,t}$ if the rating of a neighbor $u \in neighbors_n$ is used but transform that value to a binary scale:

$$\bar{r}_{impl_t}^{(u)} = \begin{cases} 1 & if\ r_{impl_t}^{(u)} \geq 1 \\ 0 & if\ r_{impl_t}^{(u)} = 0 \end{cases} \tag{1.30}$$

This is done because it to some extent does not matter *how much* posts a neighbor has made in a thread (this information is only used when calculating similarities between users according to equation (1.28)). The important thing is the vote of the neighbor for that thread as it being a relevant thread to user n.

One algorithm, called *question-answer*, uses a very discussion board specific definition of similarity between two users. The basic idea is that the order of posts in threads implicates social relationships between the authors, i.e. the social relationship between author A and author B is the higher the more often they reply to each other *directly* within threads. Though there does not have to be an explicit friendship relation in the social network, two users that reply to each other above-average in the discussion board arguably *have* at least some kind of social relationship – be it a negative or a positive one.

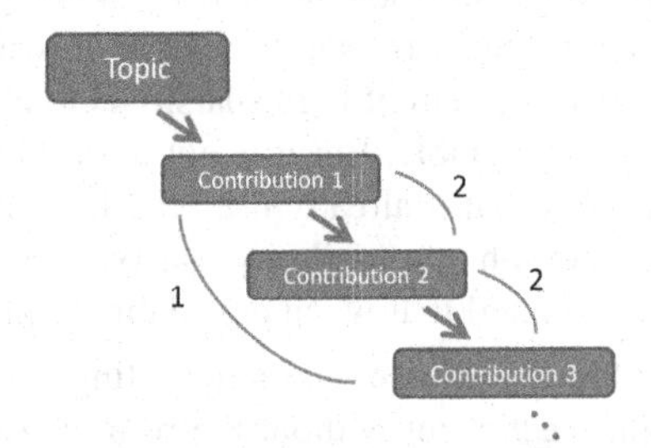

Fig. 1.3 Visualization of basic idea of *question-answer* algorithm

Based on this notion the *question-answer* algorithm redefines the similarity function:

$$\begin{aligned} sim_{n,u} = 2 \cdot |\{(p^{(1)}, p^{(2)}),\ p^{(1)}, p^{(2)} \in P :\ p^{(1)}_{author} = n \\ \wedge\ p^{(2)}_{author} = u \wedge p^{(1)}_{thread} = p^{(2)}_{thread} \\ \wedge\ p^{(2)}_{rank} - p^{(1)}_{rank} = 1\}| \\ + |\{(p^{(1)}, p^{(2)}),\ p^{(1)}, p^{(2)} \in P :\ p^{(1)}_{author} = n \\ \wedge\ p^{(2)}_{author} = u \wedge p^{(1)}_{thread} = p^{(2)}_{thread} \\ \wedge\ p^{(2)}_{rank} - p^{(1)}_{rank} = 2\}| \end{aligned} \tag{1.31}$$

A visualization of the idea can be seen in Figure 1.3. The numbers in the figure represent the differently weighted summands from the definition above if post 1 and 3 have been written by author n and post 2 by author u. *question answer* works with implicit social relations, i.e. social relations that were not stated explicitly by the users. Hence, one could think this is a social filtering approach. However this approach is related more closely to the standard collaborative filtering approach if you compare the definition of similarities (cp. equations (1.28) and (1.31)).

As in *Experiment One* two naive algorithms were implemented in *Experiment Two*. While *random* is the same as in *Experiment One*, the second one is specific for *Experiment Two*. The naive *popularity algorithm* is based on the popularity $Popularity^{(t)} = |P_t|$ of a thread:

$$pred_{n,t} = \frac{Popularity^{(t)}}{|P|} \tag{1.32}$$

Regarding the performance measures *Experiment Two* also defines f-measure and MAE:

$$\begin{aligned}
T_{relevant} &= \bigcup_{n \in N} T^{(n)}_{relevant} \\
T_{accurate} &= \bigcup_{n \in N} \{t \in T^{(n)}_{selected} \mid t \in T^{(n)}_{relevant}\} \\
T_{inaccurate} &= \bigcup_{n \in N} \{t \in T^{(n)}_{selected} \mid t \notin T^{(n)}_{relevant}\} \\
Precision &= \frac{|T_{accurate}|}{|T_{accurate}| + |T_{inaccurate}|} \\
Recall &= \frac{|T_{accurate}|}{|T_{relevant}|} \\
F &= \frac{2 \cdot Precision \cdot Recall}{Precision + Recall}
\end{aligned} \tag{1.33}$$

$$MAE = \frac{1}{\sum_{n \in N} |T^{(n)}_{test}|} \cdot \sum_{n \in N} \sum_{t \in T^{(n)}_{test}} |pred_{n,t} - \bar{r}^{(n)}_{impl_t}| \tag{1.34}$$

We now give a list of the implemented algorithms that were tested in *Experiment One*:

- **random**
  - *neighborhood:* n/a
  - *similarity matrix:* n/a
  - *calculation of prediction:* cp. (1.21)
- **popularity**
  - *neighborhood:* n/a
  - *similarity matrix:* n/a
  - *calculation of prediction:* cp. (1.32)
- **standard collaborative**
  - *neighborhood:* via similarity matrix (cp. (1.29))
  - *similarity matrix:* similar active threads (cp. (1.28))
  - *calculation of prediction:* consideration of similarity weights (cp. (1.13))
- **social 1-level w/o similarity**
  - *neighborhood:* 1st level of friendship (cp. (1.18))
  - *similarity matrix:* n/a
  - *calculation of prediction:* simple average (cp. (1.11))

- **social 2-level w/o similarity**
  neighborhood: 2st level of friendship (cp. (1.19))
  similarity matrix: n/a
  calculation of prediction: simple average (cp. (1.11))
- **social simple clique**
  neighborhood: cliques (cp. (1.20))
  similarity matrix: n/a
  calculation of prediction: simple average (cp. (1.11))
- **question-answer**
  neighborhood: via similarity matrix (cp. (1.29))
  similarity matrix: implicit social relations (cp. (1.31))
  calculation of prediction: consideration of similarity weights (cp. (1.13))

The social filtering algorithms *social 1-level w/o similarity*, *social 2-level w/o similarity* and *social simple clique* do not consider similarity values because of the very sparse rating vectors in *Experiment Two*. Hence $sim_{n,u}$ would have often been zero so that the social filtering approaches would have been disadvantaged (*standard collaborative* and *question-answer* do not have the same problem because neighborhood is defined on basis of the similarity matrix). It should be also stated that the average ratings were not considered in the case of *standard collaborative* and *question-answer* because this would be senseless with implicit ratings.

The results of the different algorithms with regard to the peformance measures are presented in Fig. 1.4.

Discussion of Experiment Two

First of all it is apparent that all algorithms perform better than the *random* algorithm (in contrast to *Experiment One*) but this was to expect due to the test procedure it is not even possible to perform worse than *random*: z items have to be selected by an algorithm *in any case*, i.e. in the worst case this happens on a random basis but this is exactly what *random* does.

The f-measure in *Experiment Two* is more interesting than the MAE. After all the important thing is *which* items are selected by the algorithms rather than an exact prediction on an artificial binary scale. The big difference between *random* and the other algorithms with respect to the MAE is caused by an effect that makes predictions near 0 more probable for the other algorithms which occurs naturally more often with higher v.

Because of the fact that any algorithm *has* to perform better than *random* we should focus on the second naive algorithm, *popularity*, as a benchmark for the other algorithms. In comparison with *popularity* it becomes obvious that only two algorithms perform better than this naive algorithm, both of which are no social filtering approaches, *standard collaborative* and *question-answer*. The similarity of these both algorithms was already explained above in the text when introducing

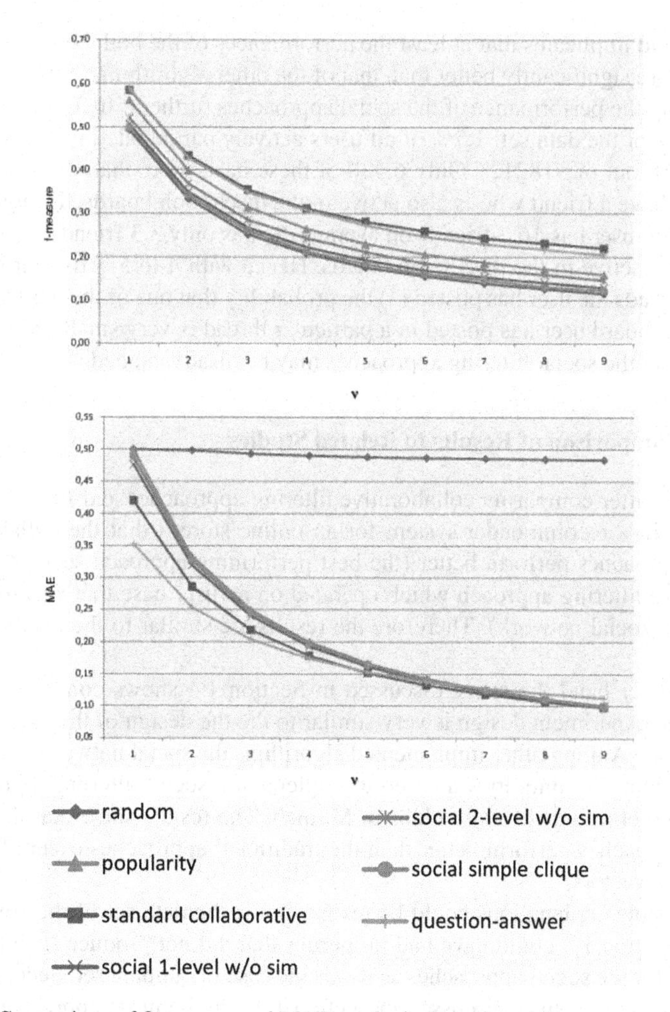

Fig. 1.4 Comparison of f-measure and MAE results in Experiment Two

question-answer. The differences between the social filtering algorithms are minor ones (with *social 2-level w/o similarity* performing slightly better than the rest). In the end we find that in this test scenario social filtering approaches are clearly not able to build up any intelligent internal model that allows for better recommendations than a very naive approach without any intelligent model.

The differences between the different algorithms are pretty small. The reason for this is that probability plays a prominent role due to the design of the test procedure (for $v = 9$ only one item out of 10 items is a relevant one) which makes high demands on any recommender system. According to that the differences of the algorithms compared to the *random* algorithm were expected to be small, which on the other hand implicates that at least the performances of the both best performing algorithms are significantly better than that of the other algorithms.

Regarding the performance of the social approaches further difficulties exist due to properties of the data set: 1.7% of all users actively participate in the discussion boards (9018 out of 518212). Only 6.430 of these have more than one friend and only 4.248 have a friend who is also active in the discussion boards. Though a discussion board user has 16.3 friends on average, he has only 3.3 friends on average, who are also active in the discussion boards. Hence with 4 to 5 active threads per user (i.e. threads the user has posted in) the probability that one of the 3.3 friends of a discussion board user has posted in a particular thread is very small. With such a sparse dataset the social filtering approaches may be disadvantaged.

1.5.1.3 Comparison of Results to Related Studies

[72] states – after comparing collaborative filtering approaches and social filtering approaches in a recommender system for an online store – that the collaborative filtering approaches perform better (the best performing approach seemed to be a collaborative filtering approach which operated on an user base that was interconnected via a social network). Therefore the results are similar to the results of our experiments.

On the other hand the study discussed in Section 1.4 shows contrary findings although the experiment design is very similar to the the design of the experiments in this section. Among other implemented algorithms the social network of users of a Munich online community was used to implement a social filtering approach in order to predict ratings of night clubs in Munich. The results show that the social filtering approaches perform better than the traditional approaches except for very sparse rating vectors.

Two possible explanations could be given: firstly, the data set of the study discussed in Section 1.4 could have had properties that did not produce such big disadvantages for the social approaches as it was the case in our data set. Secondly, the type of items in the study discussed in Section 1.4 may be more appropriate for a social filtering approach than the type of items of this study. The rating of a night club is arguably correlated with the likings of your friends. Moreover the community of the study discussed in Section 1.4 is a Munich-based one (so the chance that two cyber-friends know each other in reality is rather high, hence they may also go clubbing together) while the discussion board users of our study come from all quarters of Germany and many probably do not know each other personally.

1.5.2 Future Prospects

Concluding, it is difficult to tell how strong the performance of the social approaches is actually influenced by the slightly disadvantageous properties of the data set in this study. This is due to the existence of several other possible influencing factors regarding the field of application. The study discussed in Section 1.4, the study

[72] and the study in this section state different results concerning social filtering approaches and it is conceivable that the field of application plays a prominent role for the performance of the social approaches because the type of rated objects, the users of the system, their social connections and the type of ratings differ in all three studies.

Apparently a comparison of the results is therefore difficult. In order to be able to evaluate the potential of social recommenders in different fields of applications further studies must be made. Ideally disadvantageous properties of the data set as in this study should be avoided as far as possible in order to eliminate at least this influencing factor.

A concluding study could then eventually make an attempt to identify the actual influencing factors for the performance of social filtering approaches by categorizing the fields of applications of several studies with regard to different influencing factors such as the type of rated objects, the users of the system, and so on. After that the analysis of each single influencing factor could offer valuable clues to the relevance of each influencing factor.

This could be done in a visualization like in Fig. 1.5 where we have shown a fictional possible correlation between the average age of system users and f-measure results of social approaches in the respective study. A single point in the diagram represents one study with a particular field of application and a particular average age of its system users. The insertion of a trend line could make a possible correlation between the influencing factor and the performance of social approaches visible.

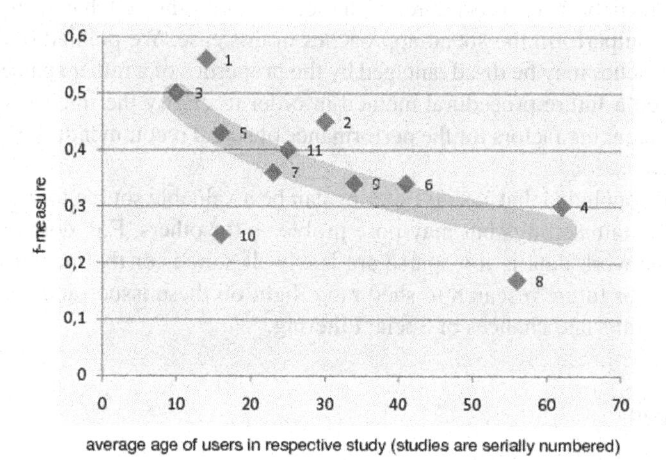

Fig. 1.5 *Fictional* correlation between the average age of users and social filtering f-measure results of different studies

1.6 Conclusion and Future Work

This chapter focused on Social Recommender Systems. After a short overview of existing recommendation approaches, we briefly reviewed related work on Social Recommender Systems. Among the various interpretations of the term we focused on approaches using social subnetworks as rating neighborhoods (Social Filtering). We compared two studies using Social Filtering. The first study, Social Filtering performed comparably well or better than traditional CF approaches in a taste related domain. Besides the performance, the concept has several advantages. It e.g. avoids traditional cold start problems. Via using the social network as a long term social recommendation context, which was explicitly stated by the users and amalgamates an enormous amount of social knowledge and attitude in a lean data-structure, even more radical recommendations may be successful. These horizon broadening recommendations, which may not reflect the user's hitherto existing or system represented taste are much more likely to be accepted if originating from a user's social neighborhood compared to an (anonymous) rating based neighborhood, especially in non-factual, taste related domains.

We compared these findings with a second empirical study that aimed at investigating Social Filtering for discussion boards. We used a data set that was downloaded from a German social networking platform and included user profiles, friendship relations and discussion board data. Several different variations of traditional collaborative filtering algorithms as well as social filtering approaches were implemented and tested in two separate test procedures which used different types of ratings for the investigation of different application scenarios.

The results in both test procedures of the second study showed that the traditional approaches outperform the social approaches in this case. We pointed out that the social approaches may be disadvantaged by the properties of a rather sparse data set and proposed a future procedural method in order to clarify the impact of several possible influencing factors for the performance of social recommenders in different fields of application.

It can be concluded that Social Filtering can be a valuable source for recommendations in certain settings but may pose problems for others. E.g. domains where the social network data is too sparse are less well suited for the method. It is an opportunity for future research to shed more light on these issues and continue to investigate limits and chances of Social Filtering.

References

1. Abel, F., Bittencourt, I.I., Henze, N., Krause, D., Vassileva, J.: A Rule-Based Recommender System for Online Discussion Forums. In: Nejdl, W., Kay, J., Pu, P., Herder, E. (eds.) AH 2008. LNCS, vol. 5149, pp. 12–21. Springer, Heidelberg (2008)
2. Adomavicius, G., Tuzhilin, A.: Toward the next generation of recommender systems: A survey of the state-of-the-art and possible extensions. IEEE Transactions on Knowledge And Data Engineering 17(6), 734–749 (2005)

3. Adomavicius, G., Sankaranarayanan, R., Sen, S., Tuzhilin, A.: Incorporating contextual information in recommender systems using a multidimensional approach. ACM Trans. Inf. Syst. 23, 103–145 (2005)
4. Amazon, http://www.amazon.de (url accessed January 2011)
5. Baeza-Yates, R., Ribeiro-Neto, B.: Modern information retrieval. Addison-Wesely (1999)
6. Balabanović, M., Shoham, Y.: Fab: content-based, collaborative recommendation. Communications of the ACM 40(3), 66–72 (1997)
7. Belkin, N.J., Croft, W.B.: Information filtering and information retrieval. Communications of the ACM 35(12), 29–38 (1992)
8. Billsus, D., Pazzani, M.J.: Learning collaborative information filters. In: Proceedings of the Fifteenth International Conference on Machine Learning, vol. 54 (1998)
9. Billsus, D., Pazzani, M.J.: User modeling for adaptive news access. User Modeling and User-Adapted Interaction 10(2), 147–180 (2000)
10. Birnkammerer, S., Woerndl, W., Groh, G.: Recommending for groups in decentralized collaborative filtering. Technical report, TU Muenchen (2009)
11. Bonhard, P., Sasse, M.A.: 'knowing me, knowing you' – using profiles and social networking to improve recommender systems. BT Technology Journal 24(3), 84–98 (2006)
12. Bortz, J.: Statistik. Springer (2005)
13. Breese, J.S., Heckerman, D., Kadie, C.: Empirical analysis of predictive algorithms for collaborative filtering. In: Proceedings of the 14th conference on Uncertainty in Artificial Intelligence, pp. 43–52. Morgan Kaufmann (1998)
14. Brocco, M., Asikin, Y.A., Woerndl, W.: Case-based team recommendation. In: Proc. of the 2nd International Conference on Social Informatics, SocInfo 2010 (2010)
15. Brocco, M., Groh, G.: Team recommendation in open innovation networks. In: Proc. Third ACM Conference on Recommender Systems, RecSys 2009 (2009)
16. Brocco, M., Groh, G.: A meta model for team recommendations in open innovation networks. In: Short Paper, Proc. Third ACM Conference on Recommender Systems (RecSys 2009), NY, USA (2009)
17. Brocco, M., Groh, G., Forster, F.: A meta model for team recommendations. In: Proc. SocInfo 2010, Laxenburg, Austria (2010)
18. Brocco, M., Groh, G., Kern, C.: On the influence of social factors on team recommendations. In: Second International Workshop on Modeling, Managing and Mining of Evolving Social Networks (M3SN), Co-located with IEEE ICDE 2010, Long Beach, USA (2010)
19. Burke, R.: Hybrid recommender systems: Survey and experiments. User Modeling and User-Adapted Interaction 12(4), 331–370 (2002)
20. Burke, R.: Hybrid Web Recommender Systems. In: Brusilovsky, P., Kobsa, A., Nejdl, W. (eds.) Adaptive Web 2007. LNCS, vol. 4321, pp. 377–408. Springer, Heidelberg (2007)
21. Chen, J., Geyer, W., Dugan, C., Muller, M., Guy, I.: Make new friends, but keep the old: recommending people on social networking sites. In: Proceedings of the 27th International Conference on Human Factors in Computing Systems, CHI 2009, pp. 201–210. ACM Press (2009)
22. Crossen, A., Budzik, J., Hammond, K.J.: Flytrap: intelligent group music recommendation. In: Proceedings of the 7th International Conference on Intelligent User Interfaces, IU 2002, pp. 184–185 (2002)
23. Cunningham, P.: A taxonomy of similarity mechanisms for case-based reasoning. IEEE Transactions on Knowledge and Data Engineering, 1532–1543 (2008)
24. Deshpande, M., Karypis, G.: Item-based top-n recommendation algorithms. ACM Transactions on Information Systems (TOIS) 22(1), 143–177 (2004)

25. Dey, A.K.: Understanding and using context. Personal and ubiquitous computing 5(1), 4–7 (2001)
26. Facebook (2011), http://www.facebook.com (url accessed January 2011)
27. Faerber, F., Weitzel, T., Keim, T.: An automated recommendation approach to selection in personnel recruitment. In: Proceedings of the 2003 Americas Conference on Information Systems (2003)
28. Felfernig, A., Burke, R.: Constraint-based recommender systems: technologies and research issues. In: Proceedings of the 10th International Conference on Electronic Commerce, ICEC 2008, pp. 3:1–3:10 (2008)
29. Filmtipset (2011), http://www.filmtipset.se (url accessed January 2011)
30. Goren-Bar, D., Glinansky, O.: Fit-recommending tv programs to family members. Computers & Graphics 28(2), 149–156 (2004)
31. Groh, G.: Groups and group-instantiations in mobile communities–detection, modeling and applications. In: Proceedings of the International Conference on Weblogs and Social Media, Citeseer (2007)
32. Groh, G., Ehmig, C.: Recommendations in taste related domains: Collaborative filtering vs. social filtering. In: Proc. ACM Group 2007, pp. 127–136 (2007)
33. Groh, G., Daubmeier, P.: State of the art in mobile social networking on the web. TU-Muenchen, Faculty for Informatics, Technical Report, TUM-I1014 (2010)
34. Groh, G., Lehmann, A., Reimers, J., Friess, R., Schwarz, L.: Detecting social situations from interaction geometry. In: Proc. IEEE SocialCom (2010)
35. Herlocker, J.L., Konstan, J.A., Terveen, L.G., Riedl, J.T.: Evaluating collaborative filtering recommender systems. ACM Transactions on Information Systems (TOIS) 22, 5–53 (2004)
36. Herlocker, J.L., Konstan, J.A., Borchers, A., Riedl, J.: An algorithmic framework for performing collaborative filtering. In: Proceedings of the 22nd annual international ACM SIGIR conference on Research and development in information retrieval, SIGIR 1999, pp. 230–237. ACM (1999)
37. Herlocker, J.L., Konstan, J.A., Riedl, J.T.: An empirical analysis of design choices in neighborhood-based collaborative filtering systems. Information Retrieval 5, 287–310 (2002)
38. Jameson, A.: More than the sum of its members: challenges for group recommender systems. In: Proceedings of the Working Conference on Advanced Visual Interfaces, pp. 48–54. ACM (2004)
39. Jameson, A., Smith, B.: Recommendation to Groups. Springer (2007)
40. Jameson, A., Baldes, S., Kleinbauer, T.: Enhancing mutual awareness in group recommender systems. In: Bamshad Mobasher and Sarabjot (2003)
41. Jameson, A., Baldes, S., Kleinbauer, T.: Two methods for enhancing mutual awareness in a group recommender system. In: Proceedings of the International Working Conference on Advanced Visual Interfaces (2004)
42. Joachims, T.: Learning to classify text using support vector machines: Methods, theory, and algorithms. Computational Linguistics 29(4), 656–664 (2002)
43. Kosub, S.: Local Density. In: Brandes, U., Erlebach, T. (eds.) Network Analysis. LNCS, vol. 3418, pp. 112–142. Springer, Heidelberg (2005)
44. Lappas, T., Liu, K., Terzi, E.: Finding a team of experts in social networks. In: Proceedings of the 15th ACM SIGKDD International Conference on Knowledge Discovery and Data Mining, pp. 467–476. ACM (2009)
45. Liben-Nowell, D., Novak, J., Kumar, R., Raghavan, P., Tomkins, A.: Geographic routing in social networks. Proceedings of the National Academy of Sciences of the United States of America 102(33), 11623–11628 (2005)

46. Lin, C.Y., Ehrlich, K., Griffiths-Fisher, V.: Searching for experts in the enterprise: combining text and social network analysis. In: Proceedings of the 2007 International ACM Conference on Supporting Group Work, Group 2007, pp. 117–126. ACM (2007)
47. Lin, C.Y., Ehrlich, K., Griffiths-Fisher, V., Desforges, C.: Smallblue: People mining for expertise search. IEEE Multimedia 15, 78–84 (2008)
48. LinkedIn (2011), `http://www.linkedin.com` (url accessed January 2011)
49. Lokalisten (2011), `http://www.lokalisten.de` (url accessed January 2011)
50. Malinowski, J., Weitzel, T., Keim, T., Wendt, O.: Decision support for team building: incorporating trust into a recommender-based approach. In: Proceedings of the 9th Pacific Asia Conference on Information Systems (PACIS 2005), Bangkok (2005)
51. Massa, P., Avesani, P.: Trust-aware collaborative filtering for recommender systems. In: In Proc. of Federated Int. Conference On The Move to Meaningful Internet: CoopIS, DOA, ODBASE, pp. 492–508 (2004)
52. Masthoff, J.: Group modeling: Selecting a sequence of television items to suit a group of viewers. User Modeling and User-Adapted Interaction (2004)
53. McSherry, D.: Explanation in recommender systems. Artificial Intelligence Review 24(2), 179–197 (2005)
54. Montaner, M., Lopez, B., de la Rosa, J.L.: A taxonomy of recommender agents on the internet. Artificial Intelligence Review 19(4), 285–330
55. O'Connor, M., Cosley, D., Konstan, J.A., Riedl, J.: Polylens: a recommender system for groups of users. In: Proceedings of the seventh conference on European Conference on Computer Supported Cooperative Work (2001)
56. Pazzani, M.J., Billsus, D.: Content-based recommendation systems. In: Brusilovsky, P., Kobsa, A., Nejdl, W. (eds.) Adaptive Web 2007. LNCS, vol. 4321, pp. 325–341. Springer, Heidelberg (2007)
57. Pennock, D.M., Horvitz, E., Lawrence, S., Giles, C.L.: Collaborative filtering by personality diagnosis: A hybrid memory- and model-based approach. In: Proc. 16th. Ann. Conf. on Uncertainty in AI (UAI 2000), pp. 473–480. Morgan Kaufmann (2000)
58. Pennock, D.M., Horvitz, E., Lee Giles, C.: Social choice theory and recommender systems: Analysis of the axiomatic foundations of collaborative filtering. In: Proceedings of the Seventeenth National Conference on Artificial Intelligence and Twelfth Conference on Innovative Applications of Artificial Intelligence, pp. 729–734 (2000)
59. Pennock, D.M., Horvitz, E., Lawrence, S., Lee Giles, C.: Collaborative filtering by personality diagnosis: A hybrid memory and model-based approach. In: Proceedings of the 16th Conference on Uncertainty in Artificial Intelligence, UAI 2000, pp. 473–480 (2000)
60. Resnick, P., Iacovou, N., Suchak, M., Berstrom, P., Riedl, J.: Grouplens: An open architecture for collaborative filtering of netnews. In: Proceedings of the 1994 ACM Conference on Computer Supported Cooperative Work, pp. 175–186. ACM Press (1994)
61. Richardson, M., Agrawal, R., Domingos, P.: Trust management for the semantic web. In: Proceedings of the Second International Semantic Web Conference, pp. 351–368 (2003)
62. Said, A., de Luca, E.W., Albayrak, S.: How social relationships affect user similarities. In: Guy, I., Chen, L., Zhou, M.X. (eds.) Proc. of 2010 Workshop on Social Recommender Systems (2010)
63. Sarwar, B., Karypis, G., Konstan, J., Riedl, J.: Item-based collaborative filtering recommendation algorithms. In: Proceedings of the 10th International WWW Conference, pp. 285–295. ACM (2001)
64. Sarwar, B.M., Karypis, G., Konstan, J.A., Riedl, J.T.: Application of dimensionality reduction in recommender system – a case study. In: ACM WEBKDD Workshop (2000)
65. Schafer, J.B., Frankowski, D., Herlocker, J., Sen, S.: Collaborative filtering recommender systems. In: Brusilovsky, P., Kobsa, A., Nejdl, W. (eds.) Adaptive Web 2007. LNCS, vol. 4321, pp. 291–324. Springer, Heidelberg (2007)

66. Sinha, R., Swearingen, K.: Comparing recommendations made by online systems and friends. In: Proceedings of the DELOS-NSF Workshop on Personalization and Recommender Systems in Digital Libraries (2001)
67. Terveen, L., McDonald, D.W.: Social matching: A framework and research agenda. ACM Trans. Comput.-Hum. Interact. 12(3), 401–434 (2005)
68. Woerndl, W., Groh, G., Hristov, A.: Individual and social recommendations for mobile semantic personal information management (2009)
69. Woerndl, W., Muehe, H., Prinz, V.: Decentral item-based collaborative filtering for recommending images on mobile devices. In: IEEE International Conference on Mobile Data Management: Systems, Services and Middleware, pp. 608–613 (2009)
70. Yaniv, I.: Receiving other people's advice: Influence and benefit. Organizational Behavior and Human Decision Processes 93 (2004)
71. Zhang, Y., Callan, J., Minka, T.: Novelty and redundancy detection in adaptive filtering. In: Proceedings of the 25th Annual International ACM SIGIR Conference on Research and Development in Information Retrieval, pp. 81–88. ACM Press (2002)
72. Zheng, R., Provost, F., Ghose, A.: Social network collaborative filtering. In: CeDER-07-04, CeDER Working Papers. New York University (2007)

Chapter 2
Legal Aspects of Recommender Systems in the Web 2.0: Trust, Liability and Social Networking

Teresa Rodríguez de las Heras Ballell

Abstract. Along with an increasing sophistication of products, services and contents migrating to the digital space, the Social Web has revolutionized the architecture of social relationships. In a social networking environment, the traditional legal analysis is challenged by new behavioral patterns likely to evoke certain privacy concerns. On the one hand, users are actively involved in the provision of relevant information intended to fuel personalization strategies and recommender systems. Therefore, the recommending activity becomes decentralized and users go from passive observer to active recommendation provider. On the other hand, the intense participation in social networks represents an infinite source of valuable information. Since like-minded individuals assemble and share interests and preferences, regular and repetitive social patterns in searching, browsing, selecting or buying offer a wealth of (shared) data about users and represent a valuable source of information. Both assumptions, linked to the Social Web, call for exploring the need to redefine the legal approach to recommender systems under new coordinates.

This paper aims to examine the development of recommender systems, considering the new patterns of social behavior, and tackle the emerging legal concerns aroused by the collection and process of personal and social information and the legal consequences of devising a decentralized recommendation model. As a conclusion, a model likely to achieve a balance between privacy risks and personalization advantages is proposed. Service providers are encouraged to devise technical and organizational measures intended to minimize data collection and guarantee a free and well-informed user consent giving by implementing opt-out mechanisms and deploying ambitious transparency and disclosure policies, enabling individuals to build their digital ego on a sound consent-directed basis.

Teresa Rodríguez de las Heras Ballell
University Carlos III of Madrid
e-mail: teresa.rodriguezdelasheras@uc3m.es

J.J. Pazos Arias et al.: Recommender Systems for the Social Web, ISRL 32, pp. 43–62.
springerlink.com

2.1 Personalization and Recommendation in the Social Web: Generating Trust

The increasing sophistication of products, services and contents migrating to the digital space [3] is revealing two outstanding features of the digital economy that are becoming more and more accentuated last years.

On the one hand, the assessment that, as information continues to grow at an exponential growth and is becoming more and more readily accessible, the scarcest asset in the digital economy proves to be reliability. Users must deal with an "informative exuberance" in the absence of tools for verifying and gauging its reliability. Information is indisputably the sap of social and economic relationships in modern economies, but only reliable and precise information can create real value. Otherwise, a flood of untruthful and inaccurate information distorts the decision-making process, misplaces confidence and increases the risks specifically involved in a given transaction. Therefore, trust-creating mechanisms are becoming indispensable for an efficient functioning of the information market [13], encumbered with significant information asymmetries. Among them, reputational intermediaries [17, 14, 5] play a remarkable role in reinforcing confidence while reducing, at the same time, the costs of collecting, processing and verifying the needed information to assess trustworthiness. Accordingly, the increasing complexity of the digital world is urgently demanding the provision of new intermediating services and the participation of novel and original intermediaries.

Far from the simple picture that offers the basic distinction between access infrastructure providers and content providers, a three-layer model for on-line intermediation is proposed. Intermediaries are to provide accessibility, visibility and credibility [6, 7]. The smooth running of Internet does not only depend on the technical infrastructure ensuring accessibility, but requires as well two further infrastructures aimed at supplying visibility-enhancing tools and creating credibility references. The second generation of e-commerce law presumes accessibility and then focuses attention on improving visibility and enhancing trust.

On the other hand, the statement that, contrary to expectations, e-business models are not basing their strengths on the provision of mass and standardized services, but on the undertaking of increasingly sophisticated personalization-based strategies. Instead of exploiting the eventual opportunities linked to the potentially worldwide extension of a monotonous and uniform target demand, a wide range of rich personalizing strategies is being implemented as a powerful commercial appeal. Businesses have become aware of the extraordinary value of information when duly managed to fuel successful personalization-based strategies, recommender systems, behavioral advertising programs and context-aware services. As a result, providers are able to tailor their customized products that are more proactive and sensitive to users' preferences, as well as gain competitive advantages by distinguishing their offer from their rivals' ones. Users, moreover, benefit from services and products meeting their real interests and from a variety of tools assisting them in browsing through the sea of information goods.

Furthermore, apart from underlying successful customization-based business strategies, a personalization-oriented approach is urgently demanded as an inspiring guideline in the devising of public information policies, transparency and disclosure duties and customer service programs. A centralized, unidirectional and reactive more than proactive conception of an information provision model, where the duty to supply data and the burden to request for information are placed on the user, is revealing serious inefficiencies. As soon as any exceptional situation (catastrophe, air controllers strike, natural disaster, air traffic stopped by an active volcano) requires a mass provision of information to users, citizens or customers, customer service systems paralyze and the processing of information requests become unworkable. A personalized provision of information is not only more efficient but, above all, it embodies a fairer and user-oriented information model. Governments, the civil service and companies are realizing the need to reorientate their information flows in order to enhance transparency, improve e-Government and provide better customer services. Real time information should be made readily available to addressees - users, citizens, clients - through electronic means on a personalized basis. Long lines in front of customer service points and permanently engaged phone numbers are failures of a pre-digital world. New technologies do not only speed up the pace of communications but enable a real redesign of structures and models. The supply of personalized information addressed to the interested party and the use of social networks to keep users duly and timely informed on a sharing basis are ushering public administration and businesses in a new way of understanding information services.

Personalization technology combines "ideas from user profiling, information retrieval, artificial intelligence and user interface design" enabling services that are better adapted to the learned needs and priorities of individuals and groups of users [22]. Nevertheless, in the supply of such value-added services, a huge amount of data has to be processed to feed personalizing, recommending or contextualizing strategies. Collecting and processing sensitive personal data, huge amounts of transactional data and ample behavioral information for the said purposes arouse certain legal concerns [2].

In a social networking environment, two further factors must be considered in the legal analysis. Firstly, users are actively involved in the provision of relevant information intended to fuel personalization strategies and recommender systems, for instance, by tagging items to be recommended. Therefore, the recommending activity becomes decentralized and users go from passive observer to active recommendation provider. Secondly, the intense participation in social networks represents an infinite source of valuable information. Assuming that in a community-based site, like-minded individuals assemble and share interests and preferences, regular and repetitive social patterns in searching, browsing, selecting or buying offer a wealth of (shared) data about users and represent a valuable source of information to fuel personalization and recommendation strategies. Therefore, a new dimension with a great informative potential emerges, the social networking.

Both abovementioned operative elements, linked to the Social Web, call for exploring the need to redefine the legal approach to recommender systems under new coordinates.

This paper aims to examine the development of recommender systems, considering the new patterns of social behavior, and tackle the emerging legal concerns aroused by the collection and process of personal and social information and the legal consequences of devising a decentralized recommendation model. Thus, the paper is structured as follows. Firstly, Part 2 is aimed at gaining an insight into legal challenges issued by social networking sites, as far as the Social Web with its particular properties frames the current development of recommender systems. Secondly, Part 3 tries to draw a crucial comparison involving three basic concepts linked to the supply of information-based services: mere information, simple opinions and real recommendations. Significant legal consequences stem from such a distinction. In particular, liability risks arising from the managing, the provision or the distribution of social networking services, recommender technologies and personalization-based applications are spotted and dealt with. And accordingly, several strategies to minimize such liability risks when providing recommender systems in the Social Web are proposed thereby. Part 4 traces a model aiming at balancing privacy concerns and personalization strategies, permeated by the idea of trust generation and governed by the prevalence of user consent, in devising recommender systems for the Social Web.

2.2 Understanding Social Networking Challenges

To gain a full insight into the legal issues aroused by recommender systems in the Web 2.0., the "social networking" factor must be first and carefully considered. The popularity of social networking sites epitomizes a revolutionary redefinition of social dynamics in the digital world that challenges traditional formulations of data protection and privacy principles.

The alternative then appears to be clear, but highly unsatisfactory. Either traditional privacy-enhancing principles, rules and devices are preserved under the same immutable paradigm, albeit adapted to new social patterns, or it is accepted that privacy expectations -in particular, "digital natives' privacy expectations"- prove to be changing [1] [12, 21, 19], and concepts and goals must be redefined or evolved

[1] In fact, case law is holding that a voluntary disclosure of personal information within social networks or similar digital communities or any affirmative action meaning a free and conscious willingness to reveal personal data or publicly share comments or contents may prove a free relinquishment of privacy protection on the matter. Courts show to a certain extent reluctance to recognize a reasonable expectation of privacy for materials users willingly post on the Internet without adopting any protective measures thereon. Therefore, the trend that social-networking information may be discoverable is gaining ground. Social-networking information is playing a crucial role in litigation and accordingly courts are no receptive to arguments of the parties when asserting a right of privacy. Moreover, social networks constitute helpful sources of information for lawyers to gather evidence in both criminal and civil cases.

accordingly [2]. It can be argued that contemporary standards for defining privacy should be crafted [11] to reflect the startling new features of the Social Web where users voluntarily participate in community-based spaces customizing and publishing their profiles, sharing personal information likely to impinge on third parties' privacy, and individually choosing their self-generated privacy settings.

2.2.1 Social Networking and New Privacy Expectations

Social networking is based on sharing interests, activities and contents. Consequently, ample personal data, submitted by the user on his/her own initiative, are posted on the site to be shared -according to the preselected privacy level-, but also information and contents uploaded by the user and related to a third party (photographs, videos, texts) who may be a member of the site or not. Traditional privacy regulation is mainly concerned with defining rules to protect users from the collecting and processing of personal data on an unjustified basis by companies and public institutions. Nevertheless, social networking sites are fed with personal data supplied by users on their own initiative or referred to by other users. Moreover, the notion of community, that pervades the whole conception of social networks, create the illusion of entering into a "friendly" and trustful space of intimacy where users are more willing to share personal data and act unwarily [3].

As a result, the line between service providers as authors and consumers as passive users (readers) is blurring and accordingly that between the two data-protection categories: "data subject" and "data controller".

Even the scope and extent of "personal data" [4] is challenged by the variety of contents likely to be published within the online community (images, videos, texts). Whereas a comment or an answer revealing ethnic origin or political opinion of a user is an evident sensitive data, it is debatable whether an image from which is likely to deduce religious belief or racial origin must be considered and processed accordingly as sensitive data. To the extent that the processing of any sensitive data requires prior explicit consent, the abovementioned discussion becomes relevant in the risk management strategies devised by site owners.

The posting of certain contents on the social site is usually accompanied by a tagging process likely to intensify the implications for third parties' privacy - that will increase anyway as facial recognition technologies improve - and aggravate the

[2] Article 29 Data Protection Working Party, *The Future of Privacy*, adopted on 1 December 2009, 02356/09/EN, WP 168.

[3] Report and Guidance on Privacy in Social Network Services. "Rome Memorandum", International Working Group on Data Protection in Telecommunications, 43rd meeting, 3-4 March 2008, Rome (Italy).

[4] Despite the wide and all-embracing definition of "personal data" described for the purposes of the European Community regulation on data protection: "any information relating to an identified or identifiable natural person ('data subject'); an identifiable person is one who can be identified, directly or indirectly, in particular by reference to an identification number or to one or more factors specific to his physical, physiological, mental, economic, cultural or social identity".

site owner's exposure to liability risks. Whereas the consent of the user regarding his/her personal data can be easily obtained by designing an adequate sign-up process and drafting clear terms and conditions along with an understandable privacy policy [20], the third party referred to in the uploaded contents has not had the opportunity to consent to for practical reasons and the site owner has not reasonable means to gather all these consents. Accordingly, the site owner might unwittingly be subject to unforeseeable obligations against third parties and be held liable for breaching them. Furthermore, site owners are not protected by the "safe harbor" provisions applicable to the supply of intermediary services - "mere conduit", caching, hosting, search engines, hyperlinks -, to the extent that privacy issues fall outside the "safe haven" regime.

Against such a backdrop, site owners might opt for precluding users from uploading contents referring to third parties. Such a decision would nonetheless deprive social networks of their main appeal and their very meaning. Then, site owners might decide to implement monitoring mechanisms to classify, filter and purge contents on privacy grounds. This is a technically unviable and economically unaffordable policy that would ruin the expansion and the survival of social networking sites.

Besides, the increasing popularity of a social community is usually linked to the implementation of "refer to a friend" schemes. This practice raises further legal concerns since it involves the processing of a third party's personal data (the "invited friend") by wish of the member but without the consent of the former and even unbeknown to the "invited friend" [5].

The abovementioned characteristics of social networking sites, albeit inherent in the idea of a social community and extensively accepted by users (particularly, by "digital natives") as reasonable trade-off for the advantages of digital living, are enough to evoke some legal concerns. Site owners must deal with delicate risks involving privacy issues and face up to potential liability situations.

2.2.2 Risks, Liability and Privacy-Enhancing Strategies for Social Networks

In the legal sense, social networking site owners are information society services providers under the EU e-commerce terminology. Besides, the provisions of the

[5] Should a site owner allow its users to send invitations to third parties, this practice does not fall within the prohibition on direct marketing. It may be indeed covered by the exception for personal communications. In order to apply such an exception, the social networking service provider must comply with the following criteria: no incentive is given to either sender or recipient; the provider does not select the recipients of the message; the identity of the sending user must be clearly mentioned; the sending user must know the full content of the message that will be sent on his behalf. Therefore, the practice to send invitations indiscriminately to the entire address book of a user is not allowed.

Data Protection Directive [6] apply to social network service providers insofar as they act as data controllers for the purposes of the Community rules [7]. Such a legal description does not differ from the ordinary provision of information society services in the infancy of the digital world. But the surge of social networking services entails the going on stage of users, normally considered data subjects, who might take on some of the responsibilities of a data controller. Then, in the social networking environment, along with *organizational* data controllers (service providers), *individual* data controllers (users) [15] are likely to assume certain obligations insofar as they "determine the purposes and means of the processing of personal data". Nevertheless, simply a quick look at the legal provisions on data controllers' obligations reveals the impracticability of such an application.

Considering the progressive sophistication of services, activities and contents carried on in the shift from "Web 2.0 for fun" to "Web 2.0 for productivity and services", the so-called "household exemption", that exempts from complying with the duties of a data controller an individual who processes personal data "in the course of a purely personal or household activity", will not be applicable in many cases. Notwithstanding the decreasing application of the "household exemption", the user is still entitled to benefit from other exemptions such as the exemption for journalistic purposes, artistic or literary expression [8]. But regardless of the eventual application of any exemption, the user might be liable according to civil and criminal laws with regard to third parties' rights - defamation, tort liability, intellectual rights violation, penal liability, and so on -. Site owners should take a great deal of care to provide private detection mechanism aimed at obtaining actual knowledge of the illicit character of the information, the activity or the behavior and act accordingly on an expeditious basis to remove or to disable access thereto - "safe harbor" for intermediary service providers -.

It can be concluded so far that social networking services challenge traditional privacy paradigm and, to a certain extent, question the foundations of the data protection regulation. Information, of varied nature, is provided by the user and voluntarily published on his/her own initiative. The sharing of such data is the driving force of social networking. But the social community dynamics enables users to

---

[6] Directive 95/46/EC of the European Parliament and of the Council, 24th October 1995, on the protection of individuals with regard to the processing of personal data and on the free movement of such data, published in O.J. L 281/31, dated on 23rd November 1995 (hereinafter, Data Protection Directive). Moreover, when applicable to the services provided by a social networking owner, Directive 2002/58/EC of the European Parliament and of the Council, dated on 12th July, concerning the processing of personal data and the protection of privacy in the electronic communications sector (hereinafter, Directive on privacy and electronic communications). O.J. L 201, dated on 31st July 2002.

[7] Article 2: "'controller' shall mean the natural or legal person, public authority, agency or any other body which alone or jointly with others determines the purposes and means of the processing of personal data; where the purposes and means of processing are determined by national or Community laws or regulations, the controller or the specific criteria for his nomination may be designated by national or Community law".

[8] Article 29 Data Protection Working Party, Opinion 5/2009 on online social networking, adopted on 12 June 2009, 01189/09/EN, WP 163, p. 6.

share information referring to other people, either members of the site or not: tagging a picture, rating a person, inviting a friend, listing the "people I have met/want to meet" at events. This new direction in the data provision, the overwhelming heterogeneity of shared contents and the centrifugal effect of any posted information, likely to refer to third parties' data, increase liability exposure of social networking service providers to a high degree. Hence, site owners should adopt a wide range of precautionary measures to minimize their accountability and promote privacy-friendly mechanisms. The main advisable measures are sketched out below.

Firstly, social networking service providers should ensure privacy-friendly and free of charge default settings are predetermined restricting access to pre-selected contacts. A minority of users manages to adjust default settings and when making any change to privacy settings an express consent is explicitly expressed to open their profile. Therefore, assuming that no active behavior from the average user to redefine access settings will be commonplace when signing up to the service, default settings should be the privacy-friendliest ones and "opt-out" schemes that would presume an implicit consent to extend access to the user's profile should be discarded.

Secondly, security maintenance within the social networking site rests on service providers. Therefore, they should take appropriate technical and organizational measures to prevent unauthorized processing of data and minimizing security risks. The site provider's diligence will be assessed according to the reasonableness of the precautionary measures, so adopted and deployed, in terms of cost-efficiency and technical feasibility. Nevertheless, special measures should be taken in case of minors by preventing direct marketing, asking for parental consent when required, avoiding sensitive data in the subscription form, establishing logical separation to a suitable degree between communities of minors and adults, implementing privacy-enhancing technologies, developing age verification software or deploying permanent "warning boxes" policies.

Thirdly, an important information duty falls on social networking service providers to guarantee an educated and rational decision by the user when joining the community. Apart from informing users of their identity, site owners should provide comprehensive and clear information about the purposes and different ways in which data are going to be processed. A careful drafting of the privacy policy and terms and conditions and a sensible devising of the signing-up process - presenting different available privacy schemes, explaining the practical consequences of selecting each one, and even assisting the user in the signing-up process - are therefore crucial tasks to minimize liability exposure. But, even more, service providers are obliged to provide, on a more proactive basis, adequate warning to users about privacy risks when they upload information on the community.

As providers not only inform about privacy risks to themselves but also advise users how process third parties' uploaded data likely to impinge the latter's rights and remind them to sensibly ask for the referred user's consent, a significant part of potential liability is allocated from providers to users. In other words, the service provider would have adopted thereby a diligent behavior to the fullest extent. As a mere venue, users' behavior on the site is beyond the provider's control provided that complete information and adequate warnings have been provided, and until actual knowledge of any illicit information or activity has been obtained. The rationale behind such a risk-allocation scheme is familiar and well balanced. Duties and the resulting liability in case of breach are allocated to whom is in a better position to prevent the risk. The site owner would be not capable of undertaking a workable monitoring policy considering the huge amounts of information posted by users on a voluntary basis. But users can reasonably filter, monitor and take the adequate steps to avert the encroachment on third parties' rights as far as their own user-generated content is concerned.

2.2.3 The Role of Consent and the Building of Our *Digital Ego*

The Social Web is revealing that traditional approach to privacy, personal data and personal boundaries in social relationships is being challenged. As far as more and more facets of our living are migrating to the digital space, new privacy matters have to be dealt with. Contrary to certain alarmist reactions, the digital living is not only aggravating personal exposure to privacy risks in an irremediable manner and with irreparable damages. Indeed, quite the opposite, the social networking environment yields a number of appealing opportunities to enrich and successfully manage our digital living.

Risks and menaces may be converted into strengths and opportunities provided that we are able to create and control our *digital ego*[9] on an educated and well-informed basis. Therefore, the core of the system is the user's consent, a free, well-informed and fully conscious consent. How, when and to which extent the user wants to understand, manage and share his/her privacy must be strongly and inseparably linked to the user's wishes and his/her express decision. Hence, a huge responsibility bears on site owners, social networks managers and service providers to devise privacy-friendly default settings, make available to users privacy-enhancing mechanisms, deploy opt-in strategies requiring an affirmative action by the data subject indicating the willingness to reduce the privacy level or to receive any device likely to collect personal data or behavioral information [10] and ignore accordingly implicit or presumed consent when entailing any risk for users' privacy. Such precautionary measures should be naturally reinforced where sensitive data are involved or certain groups needed of special protection (such as minors) are particularly exposed.

To the extent that social networking services enable users to upload, publish and share contents referring to other users (members of the site or not) and, even more, tag, rate or add comments thereto, the "consent element" have to be soundly enforced and must be widely assisted by additional technical and operative measures.

[9] The notion of "digital ego" has been masterly formulated by Professor Antonio RODRIGUEZ DE LAS HERAS. The author appreciates very much his helpful remarks and relevant comments on the matter.

[10] Article 29 Data Protection Working Party, Opinion 2/2010 on online behavioral advertising, adopted on 22 June 2010, 00909/10/EN, WP 171.

When possible and to the extent that technical feasibility and cost efficiency permit it, the consent of the referred person must be prior obtained. In most cases, the consent can only be gathered from the referred person subsequently and therefore it must be better managed as follows: no action required when tolerating and available claiming procedures to object. Therefore, service providers should, on the one hand, develop more and more precise search engines to locate eventual personal references, operating either on an automated basis - matching published contents against certain search criteria or a sample provided by the user - or by prior request of the user on a case-by-case basis. And, on the other hand, improve customer service's efficiency to handle users' claims on an expeditious basis.

Unlike mainstream belief, even if our privacy might be more exposed in the digital space than is supposedly in the natural world, possibilities to monitor any encroachment thereon with an unimaginably speed and certainty, to enforce our privacy model and to prevent further infringements are incomparably larger, more effective and more reliable.

2.3 Information, Opinions and Recommendations: The Added-Value of Recommender Systems

A further analysis of recommender systems in the Social Web and a more careful study of their main legal concerns require a prior conceptualization of some basic notions. Personalizing strategies, fuelled by a huge amount of data including personal data, transactional data or behavioral information, assist users in handling the information overload problem. Recommender technologies are able to contribute a great added-value to such strategies by supporting users' decision-making processes. Recommender systems operate by opinions and recommendations that help users to identify, discriminate and definitely choose products, contents or services suiting their needs and wishes more efficiently, particularly, in relation to complex items and sophisticated products. Both opinions and recommendations are naturally information-based outputs, but they contain a heavy subjective load and include qualitative nuances. As a result, their personalizing potential is extremely higher; they add greater value to a given transaction or relationship; and their ability to offer a customized response dramatically increases. In fact, should available information about the product be limited or the user's knowledge demanded to make a rational decision be reduced, recommendations play a crucial role in guiding decision-making processes [11] [16]. In this regard, in a context where information is scarce, a recommendation is capable of exerting a more forceful effect in

[11] Several factors such as cognitive circumstances (central vs. peripheral processing), situational perceptions (group membership and size), or interindividual differences (uniqueness-seeking) can account for the probability that an user chooses to rely on certain recommendations, ratings or popularity indications to guide their selections, although it is also proved that, in some cases, users struggle to make independent decisions or try persistently to depart from collective likes or more popular selections.

orientating the final decision than a full insight into the transaction particulars may have therein [12] [18, 9].

Whereas information is, or at least is expected to be, neutral and objective, opinions and recommendations express personal wishes, biases and judgments, and are based on subjective appreciations. Opinions and recommendations are not verifiable themselves, but only reliable to a certain extent of reasonableness. Nonetheless, trusting is a relationship among individuals, entities and institutions, based on two elements: a reasonable belief, supported by an acceptable level of verification, and another party's assertion of past facts, present facts, and future facts (promises), that the trusting party relies on. Therefore, trust in persons, institutions, and society emerges with proof. Accordingly, once the recommending party is proven trustworthy, there is no need to verify on a case-by-case basis its statements of facts, its recommendations or its promises [13]. Considering that, recommendation providers should struggle to generate trust by sending "trust signals" to the market. Then, users will be prone to rely on their ongoing recommendations provided that they appear to be trustworthy. As a consequence, reputation, expertise, previous experiences [13, 1], specialization or a large market share are variables to be considered by the addressee of a recommendation (the trusting party) as to support his/her reliance on the recommending party. But, trusting involves costs, benefits, and risks to both the trusted and the trusting parties [13]. Hence, a cost-benefit analysis is needed. Although costs and benefits are transferable and, therefore, will operate in aggregate, two simple statements are still credible. Whereas the risks and costs of reducing the risks to the trusting party are higher than the benefits, the party will not interact; as far as the costs to the trusted party of establishing its trustworthiness are higher than the benefits, the recommendation provider will not invest in generating trust. At the same time, recommender systems are acting in relation to third parties providing the recommended items and the rated services as trusted intermediaries, even, to a certain extent, as advisors. In doing so, they are reducing trust costs in electronic transactions and assuming a crucial role of gatekeepers or trusted third parties [6].

[12] Despite the assessment that opinions and recommendations prove to have a stronger impact on decision-making processes particularly in relation to complex products, in certain cases, the argument that the information provided must be considered a mere opinion rather than factual information has been paradoxically put forwarded as an excuse to prevent responsibility. In this regard, credit rating agencies have tried to justify their reprehensible behavior during the current economic crisis by contending that ratings are mere opinions, issued under the protection of the freedom of expression. Accordingly, investors should not blindly rely on ratings but just perhaps take them into consideration as one of the factors in the decision-making process to be subject to scrutiny and collated with further information and other opinions. Moreover, ratings would not be susceptible of being deemed right or wrong in accordance to a diligence model in gathering, processing and disclosing the collected information. Certain case law has accepted the abovementioned argument in favor of rating agencies - Jefferson County School District No. R-1 v. Moody's Investor's Services, Inc. (988 F. Supp. 1341): "(t)he bond market depends in large measure upon the free, open exchange of information concerning bond issues and the First Amendment is ultimately the best guarantor of the integrity of the bond rating system".

Therefore, recommendation-based strategies, one the one hand, require a particular cost-benefit analysis and, on the other hand, arouse some different legal concerns. In sum, alluring opportunities come up while new liability risks may arise.

2.3.1 Spotting Legal Concerns of Recommender Systems

Our analysis of recommender systems from a legal approach is based on two premises that condition the whole study and our main conclusions. Firstly, the starting point for our approach is that recommender and personalization strategies are, on the one hand, an important factor in the growth and expansion of the digital economy and provide users, with beneficial and more efficient mechanisms to manage the information overload problem. In this regard, a radical prohibitive response is meaningless. Users are able to benefit from such technologies provided that the invasion on their privacy sphere is limited to the largest extent. Therefore, our proposals must be devised to find a reasonable trade-off between privacy concerns and socio-economic advantages. That does not mean a relinquishment of either spotting legal concerns arising from the deployment of recommendation strategies or preventing service providers from encroaching on users' rights. On the contrary, such a balanced approach encourages improving precautionary measures, developing privacy-friendly applications, enhancing transparency and giving priority to the user's consent. The gravity of the system must be placed on the user and be subject to his/her free and well-informed consent. Secondly, any measure to be applied should be technologically neutral. Otherwise, legal intrusiveness would be likely to distort market functioning and hamper technological innovation and creativity in business. Moreover, legislation would risk becoming obsolete soon.

Considering the abovementioned premises as inspiring principles of our analysis, it might be well worth outlining now how recommender systems work in the Social Web. Basically the recommending process consists in gathering information from different sources and of a widely varied nature, managing the collected data and issuing a personalized recommendation intended to help the user to make a decision. From a legal perspective, the above sketched procedure encounters three eventual kinds of concerns: how data are collected and who provides them; what kind of information is gathered thereby; and, how the final recommendation is elaborated.

The Social Web infuses the three layers of the recommending process, described above, with some particular features. Firstly, the user actively interacts with applications, voluntarily expressing opinions, tagging items or rating products and services. Secondly, the growing presence of the user in social networking environments provides an infinite source of information likely to be used to build a detailed user's profile inferred from his/her social relationships. Thirdly, recommendations can be tuned up with great precision by considering contextual information.

Considering the currently developed strategies in the realm of recommender systems -namely, content-based methods, collaborative filtering and hybrid strategies [10]-, data can be collected from the own user, extrapolated from other users labeled as like-minded ones or inferred from contextual information. Assuming such a variety of situations, legal duties are diverse as well.

1. Explicit profiles: the value of consent and a new approach to adhesion contracts
   Should the user provide on a voluntary and informed basis the information to be collected and subsequently processed, either by completing an application form in the sign-up procedure or by expressing an opinion under a rating or recommendation form, it would be advisable to implement the following strategy. As far as such explicit profiles are created from personal data that data subjects themselves voluntarily provide to the service provider, a prior consent from the user is presumed. Despite accepting such an implied consent as valid, service providers should ensure that the said consent is limited to the data consciously provided by the user, offer the possibility to easily revoke his/her consent and allow the registered user to view and edit the associated profile.
   Nonetheless, the user should be provided, in addition, with an adequate notice about the specified, explicit and legitimate purposes the collected information is purported to; and accordingly, any data processing in a way incompatible with those purposes or for different purposes is prohibited. Such goals and purposes are very frequently hidden among a long relation of conditions titled "Terms of Use", "Terms of Service", or "Membership Agreement". Available at a link usually labeled "Terms of Use", "Legal Notice", "Site license", or, in the best scenario, Privacy Policy, located at the bottom of the page ('*low-traffic*' website area), terms and conditions aiming at governing the use of and access by the user to the information and services offered on the website and the collecting and processing of his/her personal data are predisposed and the user, according to the wording of the first lines of the said terms and conditions, is presumed, albeit unconscionably, to accept to abide by them whereby the mere use of such information and services. Nevertheless, the user is not called to read the terms nor even knows the existence of them. Even more, if the user realizes the relevance of reading them, a rather complex technical legal wording and a usually excessive length discourage any attempt thereto and hinder any opportunity to gain a full insight into them.
   In this regard, current business practices on the drafting of standard contractual terms should be drastically rethought [4]. On the one hand, the consent should be first retrieved on an opt-in basis and requested separately for any relevant purpose - for advertising purposes, in relation to the processing of personal data, expressing assent to the most restrictive standard terms, accepting to receive a cookie or similar tracking device -. On the other hand, the drafting and the presentation of standard terms should be completely redefined. As far as they are not negotiated by the user, are predisposed and imposed by the service provider and are drafted in a technical legal language, extremely long and usually quite complex, the presumption that the user (adherent) expresses a free and informed consent when accepting the terms, either on an explicit basis or an implicit one, is far from being realistic. Adhesion contracts are naturally useful and to a great extent an ideal scenario where all contracts were negotiated by both parties and tailored to meet parties' particular needs is, albeit desirable, totally unworkable in

modern economies. Then, our contention is that electronic means offer unimaginable possibilities to process and present information that should be better exploited. Our proposal consists in a multi-layer access to contractual terms. Along with the more formal drafting, for strictly legal purposes, of the contractual terms, a simpler version underlying the most relevant parts and stressing those conditions likely to impose limits on the user's behavior or restrict his/her rights should be easily made available. Depending on his/her knowledge, expertise or interest, the user might be able to choose a simple version of standard terms or a more complex one. Certain service providers, in particular, those managing social networking platforms (Facebook, Twitter) [13] likely to attract teenagers or minors are already following this trend towards a more user-friendly presentation and a more comprehensible wording of standard terms.

2. Collaborative filtering: social networking and the privacy sphere
   To the extent that information is gathered from other users and related to the data subject on grounds of some social patterns, shared interests or like-minded friends inferred from the user participation in social networking platforms, some privacy concerns become exacerbated.
   On the hand, it entails a crossed processing of data that includes behavioral information, social relationships, hobbies, professional contacts or certain habits. Whereas the profiling quality will surely be enhanced thereby, the intrusiveness in the user's sphere of privacy may reach an unbearable level. On the other hand, certain information provided by or related to a person is likely to be retrieved in order to build another user's profile and/or to create a recommendation addressed to another user. Then, whose consent is to be obtained? Are users willingly relinquishing their privacy when taking part into a social network? When users voluntarily share their private life, publish personal information and disclose their likes and dislikes, is the intimacy sphere getting narrower? Are the privacy expectations changing, becoming less strict?
   The guiding principles should be again to ensure that users are aware of the collecting of such data and the purposes they are going to be used to, accept those conditions and have the opportunity any time to revoke his/her consent, modify his/her profile or change his/her privacy settings. Considering the dynamics of the digital living and the massive amount of data to be gathered, the described rules can only be feasibly complied with by a prior devising of adequate technological solutions aimed at meticulously satisfying the measures related above. Therefore, opt-in strategies should be prevalent; profiles should be easily and at any time modified by the user; default settings must be privacy-friendly at the largest extent; and any behavior likely to jeopardize the user's privacy or other users' rights should be duly warned and an affirmative action from the user should be required to ensure full awareness thereof.

[13] Twitter, for instance, inserts simple and easy-to-retain tips among the terms clarifying the user's main rights and determining his/her basic duties (http://twitter.com, last visit 07/07/2011). Facebook, on its side, deploys a multi-layer strategy by displaying its policies, principles, terms and conditions and even several examples and explanations for illustrative purposes (www.facebook.com, last visit 07/07/2011).

3. Personal data, types of identifiers and the privacy sphere
   Along with data voluntarily provided by the user in the registering process or at any other stage, profiles, in particular, predictive profiles, are fuelled by information inferred from observing individual and collective user behavior, monitoring visited pages and ads viewed or clicked on, social networks he/she belongs to, or previous recommendations, ratings or comments. Tracking technologies and recommender systems based on contextual information retrieve data concerning the location of the data subject, his/her habits, age and gender, hobbies and interests, current concerns and future plans as inferred from emails or published comments, that enrich extraordinarily the user's profile. Moreover, the profile based on the analysis of the said information can be improved with aggregated data derived from the behavioral profile of other users who display similar social, transactional and behavioral patterns in other contexts. The possibilities to extrapolate the outputs are surprising.
   The question is whether such identifiers should be treated as personal data and accordingly subject to personal data legal framework.
   On the one hand, it might be well worth first noting that the concept of "personal data" for the purposes of European Privacy Directives [14] has been widely construed. According to the definition of personal data contained in Directive 95/46/EC [15]: "Personal data shall mean any information relating to an identified or identifiable natural person (...)", considering that "an identifiable person is

[14] EU Directives shall apply to data collecting, processing and transfer provided that they fall under their application scope. As far as territorial connecting factors are concerned, Article 4 of the Directive 95/46/EC and Article 3 of the Directive 2002/58/EC must be considered: Article 4: "1. Each Member State shall apply the national provisions it adopts pursuant to this Directive to the processing of personal data where: a). the processing is carried out in the context of the activities of an establishment of the controller on the territory of the Member State; when the same controller is established on the territory of several Member States, he must take the necessary measures to ensure that each of these establishments complies with the obligations laid down by the national law applicable; b). the controller is not established on the Member State's territory, but in a place where its national law applies by virtue of international public law; c). the controller is not established on Community territory and, for purposes of processing personal data makes use of equipment, automated or otherwise, situated on the territory of the said Member State, unless such equipment is used only for purposes of transit through the territory of the Community. 2. In the circumstances referred to in paragraph 1 (c), the controller must designate a representative established in the territory of that Member State, without prejudice to legal actions which could be initiated against the controller himself." Article 3: "This Directive shall apply to the processing of personal data in connection with the provision of publicly available electronic communications services in public communications networks in the Community, including public communications networks supporting data collection and identification devices."

[15] Art. 2 (a), Directive 95/46/EC of the European Parliament and of the Council, dated on 24th October 1995, on the protection of individuals with regard to the processing of personal data and on the free movement of such data, published in O.J. L 281/31, dated on 23rd November 1995 (hereinafter, Personal Data Directive).

one who can be identified, directly or indirectly, in particular by reference to an identification number or to one or more factors specific to his physical, physiological, mental, economic, cultural or social identity". Information becomes sensitive for privacy purposes as far as it is related to a natural person that is identified or may be identifiable by means of certain identifiers taking into account all the factors at stake [2]. While identification through the name is the most evident practice, other identifiers may be successfully used to single someone out. IP and e-mail addresses, or log-ins have been considered data relating to an identifiable person under certain circumstances, although such an assertion expounded in general and absolute terms is not undisputed. Therefore, even if the trend appears to be interpreting the legal notion of "personal data" as to embrace a wide range of identifiers, any data to be deemed as such should be compared with the four founding elements of the definition: "any information", "relating to", "an identified or identifiable", and "natural person". Should personal data legislation be applicable, data collecting, processing and transfer are subject to the following principles: fair and lawful data collecting means; specified, explicit and legitimate purposes, prohibiting any further data processing in a way incompatible with those purposes or for different purposes; adequacy, relevancy and minimization of data in relation to the purposes for which they are collected and/or further processed; and, accuracy and up-to-date state of data.
On the other hand, even if data are not qualified as personal data, service providers are bound to comply with certain obligations in collecting and processing data. The protection of the confidentiality of communications entails that the use of tracking technologies or any similar device enabling to monitor user's behavior, retrieve data stored in the data subject's terminal equipment or even access to location information involves an intrusion into the private sphere of users. Irrespective of the fact that the collected information is personal data within the meaning of EU Directives, the protection of the individual private sphere triggers several obligations contained in Article 5(3) of the Directive 2002/58 [16] on privacy and electronic communications. Accordingly, obtaining informed consent to lawfully store information or to gain access to information stored in the terminal equipment of a user is required. Consent prior to the data collecting is

[16] Art. 5(3) of the Directive 2002/58/EC of the European Parliament and of the Council of 12 July 2002 concerning the processing of personal data and the protection of privacy in the electronic communications sector (Directive on privacy and electronic communications) as amended by subsequent instruments (OJ L 201, 31.7.2002, p. 37): "Member States shall ensure that the storing of information, or the gaining of access to information already stored, in the terminal equipment of a subscriber or user is only allowed on condition that the subscriber or user concerned has given his or her consent, having been provided with clear and comprehensive information, in accordance with Directive 95/46/EC, inter alia, about the purposes of the processing. This shall not prevent any technical storage or access for the sole purpose of carrying out the transmission of a communication over an electronic communications network, or as strictly necessary in order for the provider of an information society service explicitly requested by the subscriber or user to provide the service. (...)".

needed even if the information collected is not considered personal data; in such a case, in addition to Article 5(3), Directive 95/46/EC will also apply. The obligation to obtain prior informed consent is to be technically articulated in two basic solutions. A transparency policy intended to duly inform users about the use of tracking mechanisms, cookies and monitoring devices, the nature of the data collected thereby and which are the purported goals of the data processing. And the devising of opt-in mechanisms, moving away from opt-out ones that do not guarantee a prior informed user consent but only an implied unconscious acceptance. Once expressly accepted by the user, to ensure that data subjects remain aware of the monitoring of his/her behavior, it would be advisable to limit in time the scope of the consent, offer the possibility to revoke the consent to being monitored and, where it is technically possible and effective, create a visible symbol of other tools reminding individuals of the monitoring and helping them to control his/her exposure to risks.

Once data have collected and processed, the last stage in the recommendation strategy is the issuing of the recommendation itself or the displaying of a selection of recommended items. Two main questions - how such an outcome is elaborated and which impact is likely to have on the decision-making process - lead us to propose a two-fold approach thereto: reliability potentialities and liability risks.

Recommender technologies, ranging from collaborative filtering or content-based filtering to more sophisticated knowledge-based recommender applications [8], entails the gathering, storing and processing of data (user preferences, wishes and needs, previous transactions, product properties and so on) to determine a recommendation. In this regard, more accurate the collected information, more diligent the data processing and trust worthier the data processor are, more reliable a recommendation becomes and more willing the user is to rely on it. At the same time, more reasonable the reliance on the recommendation is, more accountable the recommender becomes to the user for his/her (harmful) decision relying on the issued recommendation.

Firstly, it is clear that recommender systems (operators) can be negligent in the collection, interpretation and supply of information, and that can cause damage to the user relying on the resulting recommendation. Secondly, the information gathered and processed can be false or inaccurate. Thirdly, the recommendation itself can be biased, intentionally tendentious or fraudulently untrue with the ultimate aim of distorting the use decision-making process. Such considerations address the role that civil liability rules should have when applied to wrongful information in light of the new role of information in modern and future markets. Liability rules would deter the recommendation provider from carelessly collecting data, negligently processing them and being reckless, fraudulent or merely neglected in supplying recommendations.

Accordingly, recommendation providers will be prone to strengthen precautionary measures to minimize their liability exposure. Several strategies would be advisable to implement. As far as recommender systems process to a great extent information supplied by or captured from third parties (the target user, like-minded users or third parties selected on certain criteria), accuracy cannot always been guaranteed. In such cases, such restrictions should be expressly and visibly noticed to users along with a disclaimer as follows. Recommender systems would warn users that recommendations they provided are based on data, opinions, tagging or ratings stored by third parties and they are not able to vouch for their accuracy or sensibleness. In addition, recommender applications should enhance data processing in order to obtain balanced and reasonable outcomes. Besides, criteria guiding the data collecting, selection and processing and determining the resulting recommendation - qualitative criteria or quantitative ones - should be disclosed. Therefore, the user when relying on the recommendation will be fully aware of technical, procedural and substantive constrains of the result. Any attempt of the recommendation provider to conceal such constraints or restrictions might unfavourably affect the user's choice.

As a conclusion, in the trade-off between trust and liability, recommender systems should pledge a firm commitment to the implementation of a two-fold policy. On the one hand, in order to enhance the accuracy of information, they have to struggle for ensuring a meticulous collection of data, an extremely diligent processing of them and a straightforward presentation of any data restrictions and procedural constraints. On the other hand, with the aim of guaranteeing that the user's decision is free and well-informed and his/her reliance on the recommendation is reasonable, they must deploy an ambitious transparency policy and disclosure strategies.

2.4 Achieving a Balance between Privacy and Personalization in Recommender Systems: Generating Trust and Enforcing Liability Rules

The so-called Web 2.0 is ushering social living and business relationships to a new era. Far from a monotonously uniform and simply flat "global village", unstoppable trends towards the developing of personalization-based business models, recommender systems and context-based services are defining the path to be followed by business strategies, content service providers and social community-based projects. The rationale behind is meaningful, unprocessed data in bulk, albeit readily accessible, are useless; even more, are extremely likely to paralyze the decision-making process. Moreover, in absence of reliability-assessing tools, untrustworthy information is likely to misplace confidence and increase risks. The digital revolution has made, in its infancy, information readily accessible on a global and ubiquitous basis; but, individuals struggle to manage an overwhelming "information exuberance". In the second generation of digital living, efficient tools assisting users in verifying information, gauging a reasonable degree of reliability and making rational and well-informed decisions are needed. Personalization-based services, recommender technologies and other similar applications help users to choose on an educated, well-informed and efficient basis. They are able to create value and generate trust by reducing uncertainties, minimizing risks involved in transactions and overcoming asymmetries in digital relationships. Nonetheless, to the extent that personal data, behavioural information and further users identifiers have to be collected, stored and processed to fuel such strategies, some legal concerns, in particular, on privacy fields, arouse.

Against such a backdrop and considering the advisability of a trade-off between the above mentioned drawbacks and opportunities, only a model likely to achieve a balance between privacy risks and personalization advantages is to be proposed. Service providers are encouraged to devise technical and organizational measures intended to minimize data collection and guarantee a free and well-informed user consent giving by implementing opt-out mechanisms and deploying ambitious transparency and disclosure policies, enabling individuals to build their *digital ego* on a sound consent-directed basis.

References

1. Anderson, D.S.: What trust is in these times? examining the foundation of online trust. Emory Law Journal 54(3), 1441–1474 (2005)
2. De Las Heras Ballell, T. R.: Legal framework for personalization-based business models. In: Personalization of Interactive Multimedia Services: A Research and Development Perspective, pp. 3–23. Nova Publisers (2009)
3. De Las Heras Ballell, A.R.: La migración digital. Telos. Cuadernos de Comunicación, Tecnología y Sociedad 61, 4–6 (2004)
4. De Las Heras Ballell, T.R.: Terms of use, browse–wrap agreements and technological architecture: spotting possible sources of unconscionability in the digital era. Contratto e Impresa. Europa 14(2), 849–869 (2009)
5. De Las Heras Ballell, T.R.: El tercero de confianza en el suministro de información. propuesta de un modelo contractual para la sociedad de la información. Anuario de derecho civil 63(3), 1245–1284 (2010)
6. De Las Heras Ballell, T.R.: Intermediación en la red y responsabilidad civil. Revista Española de Seguros 142, 217–259 (2010)
7. De Las Heras Ballell, T.R.: La responsabilidad de los prestadores de servicios de intermediación en la red. Revista Derecho y Tecnología 11, 69–96 (2010)
8. Felferning, A., Teppan, E.: User Acceptance of Knowledge-based Recommender. In: Personalization Techniques and Recommender Systems, vol. 70. World Scientific (2008)
9. Fernández, M.A., De Las Heras Ballell, T.R.: Las agencias de "raiting" como terceros de confianzaresponsabilidad civil extracontractual y protección de la seguridad del tráfico. Revista de derecho bancario y bursátil 29(120), 141–177 (2010)
10. Fernández, Y.B., Arias, J.J.P., Solla, A.G., Cabrer, M.R., Nores, M.L.: Personalization Strategies and Semantic Reasoning: Working in tandem in Advanced Recommender Systems. In: Personalization Techniques and Recommender Systems, vol. 70, pp. 191–222. World Scientific (2008)
11. Findlay, D.: Tag! now you're really "it" what photographs on social networking sites mean for the fourth amendment. North Carolina Journal of Law and Technology 10(1), 171–202 (2009)
12. Nelson, S., Simek, J., Foltin, J.: The legal implications of social networking. Regent University Law Review 22(1), 1–34 (2009-2010)
13. Frankel, T.: Trusting and non-trusting on the internet. Boston University Law Review 81, 457–458 (2001)
14. Michael Froomkin, A.: The essential role of trusted third parties in electronic commerce. Oregon Law Review 75, 49–116 (1996)

15. Garrie, D.B., Duffy-Lewis, M., Wong, R., Gillespie, R.L.: Data protecion: the challeges facing social networking. International Law and Management Review 6, 127–152 (2010)
16. Knobloch-Westerwick, S., Sharma, N., Hansen, D.L., Alter, S.: Impact of popularity indications on readers' selective exposure to online news. Journal of Broadcasting and Electronic Media 49(3), 296–313 (2005)
17. Kraakman, R.H.: Gatekeepers: The anatomy of a third-party enforcement strateg. Journal of Law, Economics and Organization 2(1), 53–104 (1986)
18. Manns, J.D.: Rating risk after the subprime mortgage crisis: A user fee approach for rating agency accountability. North Carolina Law Review 87(1011), 1047–1056 (2009)
19. Minottie, K.: The advent of digital diaries: Implications of social networking sites for the legal profession. South Caroline Law Review 60, 1057–1064 (2008-2009)
20. Mitton, A.: Data protection and web 2.0: Whose data is it anyway? Convergence 3(1), 94–98 (2007)
21. Payne, A.C.: Twitigation: Old rules in a new world. Washburn Law Journal 49(3), 841–870 (2010)
22. Smyth, B.: Personalization-Privacy Trradeoffs in Adaptive Information Access. In: Personalization Techniques and Recommender Systems, vol. 70, pp. 3–31. World Scientific (2008)

Part II
Interoperability for Recommendation

Chapter 3
Challenges in Tag Recommendations for Collaborative Tagging Systems

Robert Jäschke, Andreas Hotho, Folke Mitzlaff, and Gerd Stumme

Abstract. Originally introduced by social bookmarking systems, collaborative tagging, or social tagging, has been widely adopted by many web-based systems like wikis, e-commerce platforms, or social networks. Collaborative tagging systems allow users to annotate resources using freely chosen keywords, so called *tags*. Those tags help users in finding/retrieving resources, discovering new resources, and navigating through the system.

The process of tagging resources is laborious. Therefore, most systems support their users by *tag recommender components* that recommend tags in a personalized way. The Discovery Challenges 2008 and 2009 of the European Conference on Machine Learning and Principles and Practice of Knowledge Discovery in Databases (ECML PKDD) tackled the problem of tag recommendations in collaborative tagging systems. Researchers were invited to test their methods in a competition on datasets from the social bookmark and publication sharing system BibSonomy. Moreover, the 2009 challenge included an online task where the recommender systems were integrated into BibSonomy and provided recommendations in real time.

Robert Jšchke
Knowledge & Data Engineering Group, University of Kassel, Wilhelmshöher Allee 73, 34121 Kassel, Germany
e-mail: jaeschke@cs.uni-kassel.de

Andreas Hotho
Data Mining and Information Retrieval Group at LS VI, University of Würzburg, Am Hubland, 97074 Würzburg, Germany
e-mail: hotho@informatik.uni-wuerzburg.de

Folke Mitzlaff
Knowledge & Data Engineering Group, University of Kassel, Wilhelmshöher Allee 73, 34121 Kassel, Germany
e-mail: mitzlaff@cs.uni-kassel.de

Gerd Stumme
Knowledge & Data Engineering Group, University of Kassel, Wilhelmshöher Allee 73, 34121 Kassel, Germany
e-mail: stumme@cs.uni-kassel.de

J.J. Pazos Arias et al.: Recommender Systems for the Social Web, ISRL 32, pp. 65–87.
springerlink.com

In this chapter we review, evaluate and summarize the submissions to the two Discovery Challenges and thus lay the groundwork for continuing research in this area.

3.1 Introduction

Collaborative tagging systems are web based systems that allow users to assign keywords – so called *tags* – to arbitrary resources. Tags are used for navigation, finding resources and serendipitous browsing and thus provide an immediate benefit for users. These systems usually include tag recommendation mechanisms easing the process of finding good tags for a resource. Delicious,[1] for instance, had a tag recommender in June 2005 at the latest,[2] BibSonomy[3] since 2006. Typically, such a recommender suggests tags to the user when she is annotating a resource. The dialog which includes these recommendations in BibSonomy, for example, is shown in Figure 3.1. In the case of BibSonomy the system offers up to five recommended tags. Recommending tags can serve various purposes, such as: increasing the chances of getting a resource annotated, reminding a user what a resource is about and consolidating the vocabulary across the users. Furthermore, as Sood et al. [41] point out, tag recommendations "fundamentally change the tagging process from generation to recognition" which requires less cognitive effort and time.

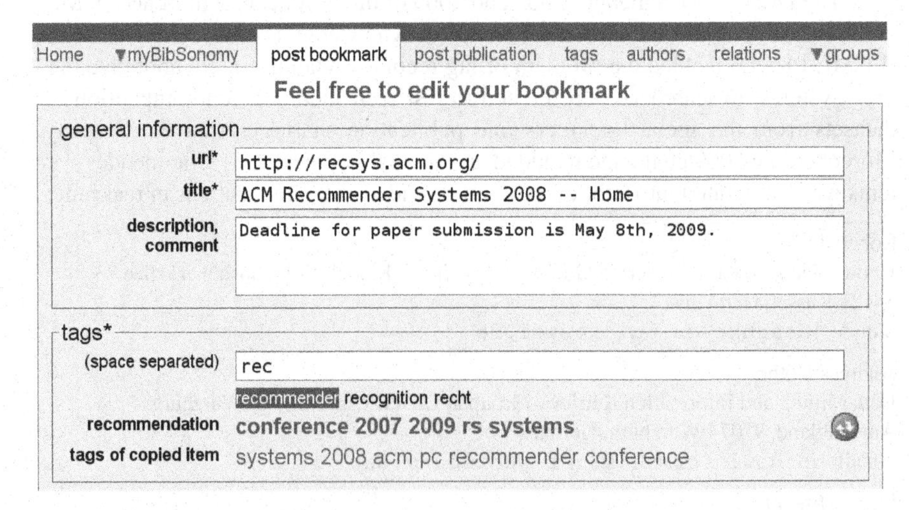

Fig. 3.1 Tag recommendations in BibSonomy during annotation of a bookmark.

[1] http://delicious.com/

[2] http://www.socio-kybernetics.net/saurierduval/archive/2005_06_01_archive.html

[3] http://www.bibsonomy.org/

In this chapter we base our investigation on the social bookmark and publication management system BibSonomy [4] which allows users to store bookmarks and publication references. We specifically focus on the tag recommender component, aiming at the analysis of all kinds of models which effectively predict the keywords a user has in mind when describing a web page (or publication). We present results for tag recommendation methods competing at two challenges we organized in the context of the ECML PKDD conference in 2008[4] and 2009[5]. The submissions to these challenges are collected in [19] and [11], resp. This paper summarizes and compares their findings.

Both challenges attracted many researchers and gave new and exciting insights. More than 150 participants registered for the mailing lists of the two challenges which enabled them to download the datasets. At the end, we received five submissions in 2008. Based on the insights of the 2008 challenge, in 2009 the tag recommendation challenge was divided into three tasks: 1) content-based, 2) graph-based, and 3) online tag recommendations. They received 21, 21, and 13 submissions, resp.

The division into three tasks is reflected in the structure of the chapter. We start with introducing the necessary fundamentals in Section 3.2 followed by our evaluation setting and a description of the datasets in Section 3.3. In Section 3.4 we present the results of the content-based tag recommendation task, Section 3.5 contains the results of the graph-based task, and Section 3.6 the results of the online task. We conclude the chapter with a discussion of lessons learnt for our specific setting, and then discuss future challenges in tag recommendation research in general.

3.2 Foundations and State of the Art

3.2.1 Tag Recommendation Problem

A folksonomy consists of sets of *users* (U), *tags* (T) and *resources* (R), connected by a ternary relation $Y \subseteq U \times T \times R$ which we call *TAS* (short for *tag assignment*) [18]. The restriction of Y to a single user is called the *personomy* of that user and contains all his tag assignments. Typically, a user u manages all tags T_{ur} he attaches to a resource r in one block which we call a *post* (u, T_{ur}, r). Thus, a post consists of a user, a resource and all tags the user has assigned to that resource.

The tag recommendation problem can be formalized as follows: Given a user u and a resource r, the goal is to predict the set T_{ur} of tags the user will assign to the resource. In the following we will depict the ordered(!) set of recommended tags by $\tilde{T}_{ur}$. Although we do not take the order of tags as the user entered them into account, the order of tags as given by the recommender plays an important role for the evaluation – following the assumption that tags that are recommended first should be the most relevant ones.

Indeed the recommendation task could be subdivided into four different tasks, depending on how much (if any) information about the user or resource or both

[4] `http://www.kde.cs.uni-kassel.de/ws/rsdc08/`

[5] `http://www.kde.cs.uni-kassel.de/ws/dc09/`

combined is available in the training data. To investigate different recommendation strategies based on the available training data used by our participants, we split the Discovery Challenge 2009 challenge into two (out of four possible) offline tasks and one online task as described below.

3.2.2 Tag Recommendation Tasks

As an analysis of the results of the 2008 Discovery Challenge showed, the *cold start problem*, i. e., the task of recommending tags for unseen users or resources, has a serious impact on the performance of certain types of recommenders. Other challenges that recommenders do not have to face in typical offline evaluations are *timeouts* and the actual exposure of the tags to the users. To address these issues, we divided the 2009 challenge into three separate recommendation tasks:

Task 1: Content-Based Tag Recommendations. The test data for this task contained posts, whose user, resource or tags were not contained in the post-core at level 2 [23] of the training data. Thus, methods which can not produce tag recommendations for new resources or are unable to suggest new tags very probably would not produce good results.

Task 2: Graph-Based Recommendations. This task was especially intended for methods relying on the graph structure of the training data only. The user, resource, and tags of each post in the test data were all contained in the training data's post-core at level 2.

Task 3: Online Tag Recommendations. The participants had to implement a recommendation service which could be called via HTTP by BibSonomy's recommender when a user posted a bookmark or publication. All participating recommenders were called on each posting process, and one of them chosen randomly to actually deliver its results to the user. We then measured the performance of the recommenders in an online setting, where timeouts were important and where we could measure which tags the user clicked.

3.2.3 State of the Art

Since the emergence of collaborative tagging systems, the topic of tag recommendations has raised quite some interest. Instead of an extensive overview on related work, we provide pointers to more elaborate surveys of state of the art (tag) recommendation methods for collaborative tagging systems. A good overview can be found in [22]. Recent works comprise [27], [34], and [40] . More generally, the state of the art on recommender systems for social bookmarking is presented in [5] and [30].

A big momentum into recommendation research brought the Netflix Prize challenge[6] held from 2007 to 2009. It was also the basis of the KDD Cup 2007 [3, 2].

[6] http://www.netflixprize.com/

The ACM RecSys conference 2010 featured a challenge on context-aware movie recommendation [37].

The ECML PKDD embraces the tradition of organizing a Discovery Challenge since 1999, allowing researchers to develop and test data mining algorithms for novel and real world datasets. The Discovery Challenge 2008 was the first that focused on recommendations and the 2009 challenge was the first offering an online competition within a real system.

3.3 Evaluation

In this section we describe the evaluation measures we used, the basics of the datasets we provided, and the baseline methods we use for comparison purposes.

3.3.1 Measures

As performance measure, we applied the F1 measure (f1m) based on precision and recall which are standard in such scenarios [17]. For each post (u, T_{ur}, r) from the test data, we compared the set $\tilde{T}_{ur}$ of recommended tags with the set T_{ur} of tags which the user has finally assigned. Then, precision and recall of a recommendation are defined as follows:

$$\mathrm{recall}(T_{ur}, \tilde{T}_{ur}) = \frac{|T_{ur} \cap \tilde{T}_{ur}|}{|T_{ur}|} \qquad \mathrm{precision}(T_{ur}, \tilde{T}_{ur}) = \frac{|T_{ur} \cap \tilde{T}_{ur}|}{|\tilde{T}_{ur}|} \; .$$

For an empty recommendation $\tilde{T}_{ur} = \emptyset$, we set $\mathrm{precision}(T_{ur}, \emptyset) = 0$. We then averaged these values over all posts from the test dataset and computed the F1 measure as $\mathrm{f1m} = 2(\mathrm{precision} \cdot \mathrm{recall})/(\mathrm{precision} + \mathrm{recall})$. Before intersecting T_{ur} with $\tilde{T}_{ur}$, we cleaned the tags in each set by ignoring the case of tags and removing all characters which are neither numbers nor letters. We ignored tags which were "empty" after normalization (i. e., they contained neither a letter nor a number). Finally, the winner was determined by the best F1 measure when regarding the *first five tags* of $\tilde{T}_{ur}$.[7]

3.3.2 Datasets

Since BibSonomy is run by our institute, we were able to provide complete snapshots of its database. BibSonomy allows users to both manage and annotate URLs (bookmarks, favorites) and BibTeX-based publication references. The snapshots thus contain both types of posts. Here we describe properties common to all of the released datasets. More specific information can be found in the corresponding

[7] In 2008 the first *ten* tags were used. For 2009 we decided to reduce the number to five, since this is more realistic from a practical point of view. In BibSonomy, only five recommended tags are shown.

sections. The datasets can be downloaded, following the procedure described at http://www.kde.cs.uni-kassel.de/bibsonomy/dumps/.

Overview. The dataset contains the dumps of several tables of the BibSonomy MySQL[8] database. Using the *mysqldump* program we dumped the tables described in Table 3.1 into the identically named files. To import those files again into a MySQL database, one needs to create the tables and load the files using the *LOAD DATA INFILE* command into the database. A post is identified by its *content_id*, resources by their *url_hash* (bookmarks) or *simhash1* (publications).

Table 3.1 The tables of the training dataset. Each table is stored as a MySQL dump in a file which has the same name as the table.

table	description
tas	fact table for tag assignments: who attached which tag to which resource (columns: user, tag, content_id, content_type, date)
bookmark	dimension table for bookmarks (columns: content_id, url_hash, url, description, extended description, date)
bibtex	dimension table for publications (columns: content_id, journal, volume, chapter, edition, month, day, booktitle, howPublished, institution, organization, publisher, address, school, series, bibtexKey, url, type, description, annote, note, pages, bKey, number, crossref, misc, bibtexAbstract, simhash0, simhash1, simhash2, entrytype, title, author, editor, year,)

Anomalies. Due to technical restrictions or unusual usage the data of BibSonomy (and probably every open system) contains various anormal fragments. Among these are spam users or posts,[9] automatically generated posts, users with very many posts, very frequent, regular or system tags and impossible dates. In 2009 we cleaned all tags as described in Section 3.3.1 and removed noisy tags (e. g., having a comma at the end) and some automatically generated tags (*imported, public, systemimported, nn, systemunfiled*). Apart from this we did not alter the data, therefore not all artifacts were removed [26]. This most closely resembles the situation of a deployed recommender system.

3.3.3 Baselines

For the offline tasks, we used four baseline methods to judge the quality of the approaches. Two of them (*all train tags known* and *all content tags known*) present an upper limit for recommendation methods that use only tags from the training data or from the content of resources. The other two methods recommend frequently used tags of the resources or user [23].

[8] http://www.mysql.com/

[9] Although BibSonomy contains a spam detection framework [4], not all spam could be removed.

All Train Tags Known. We recommend for each post from the test dataset all the tags the user assigned to it which are already contained in the training data. This is the optimum for all algorithms which use only tags from the training data. Note that the precision does not reach 100 % because the tags of some posts could not be recommended at all.

All Content Tags Known. We focus on the content of posts that is readily available at posting time and also contained in the challenge data: the URL, title, and description fields (for publication references also abstract, note, annote, series, booktitle, journal). For each post we recommend those tags we extracted from the content that the user finally assigned. This is the optimal assignment for that tag source.

Most Popular Tags By User. Given a post, we take all the tags the user assigned to resources in the training data, order them by the frequency of their usage by the user and take the most frequent tags first.

Most Popular Tags By Resource. Given a post, we take all the tags users attached to the resource in the training data, order them by the frequency of their usage and take the most frequent tags first.

3.4 Content-Based Tag Recommendations

The test data of this task might contain new users, resources, or tags. In particular, not every tag, resource, or user is already contained in the post-core at level 2 of the training data. Due to the power-law distribution of folksonomies [9], this applies to most of the nodes. Therefore, we here also analyse the results of the 2008 challenge, since there we did not divide the dataset into two parts.

3.4.1 Datasets

Both training and test data were provided as raw database dumps as described in Section 3.3.2. The dumps for the *training datasets* include all posts from BibSonomy up to and including March 31st 2008 and December 31st 2009, respectively, except more than one million posts from the user *dblp*,[10] since this user just represents a mirror of the DBLP computer science bibliography.[11] Table 3.2 shows the number of items in the datasets. We divided the statistics into bookmarks and publication references. The general format of the *test data* is as described in Section 3.3.2. Table 3.3 shows the number of items in the test datasets.

[10] `http://www.bibsonomy.org/user/dblp`

[11] `http://www.informatik.uni-trier.de/~ley/db/`

Table 3.2 The number of posts, resources, tags, and users in the training data for Task 1.

	2008 (until March 31st 2008)			2009 (until Dec. 31st 2009)		
	overall	bookmarks	publications	overall	bookmark	publications
#posts	268,692	176,147	92,545	421,928	263,004	158,924
#resources	227,764	156,059	71,705	378,378	235,328	143,050
#users	2,467	1,811	1,211	3,617	2,244	1,373
#tags	69,904	-	-	93,756	-	-

Table 3.3 The number of posts, resources, tags, and users in the test data for Task 1.

	2008 (April 1st – May 15th)			2009 (January 1st – June 30th)		
	overall	bookmark	publications	overall	bookmark	publications
#posts	59,542	16,194	43,348	43,002	16,898	26,104
#resources	59,099	15,976	43,123	40,729	15,964	24,765
#users	468	322	271	1,591	891	1,045
#tags	59,542	-	-	34,051	-	-

3.4.2 *Approaches*

Since more than 20 participants submitted results to both challenges, we here focus on the three best submissions of both 2008 and 2009. We first give a short overview on these six approaches and then provide details on preprocessing, data selection, personalization, and scoring.

3.4.2.1 Overview

The winners of the Discovery Challenge 2008 were Tatu et al. [28] using a natural language approach to generate tag recommendations. Their method includes an extensive preprocessing to clean the data. The good results mainly result from an extension of the folksonomy data with conceptual information from Wordnet [12] and from further external resources. M. Lipczak [25] was second. He developed a three step approach which utilizes words from the title expanded by a folksonomy driven lexicon, personalized by the tags of the posting user. Katakis et al. [20] came in third by considering the recommendation task as a multilabel text classification problem with tags as categories.

In 2009, Lipczak et al. [26] were the winners of Task 1. Their method – an evolution of [25] – is based on the combination of tags from the resource's title, tags assigned to the resource by other users, and tags in the user's profile. The system is composed of six recommenders and the basic idea is to augment the tags from the title by related tags extracted from two tag-tag–co-occurrence graphs and from the user's profile, and then rescore and merge them. The second best team of Task 1, Mrosek et al. [31], harvested tags from external sources and employed the full-text of web pages and PDFs. The third best team, Ju and Hwang [21], merged tags which have been earlier assigned to the resource or used by the user as well as resource descriptions by a weighting scheme.

3.4.2.2 Preprocessing

Since both training and test data were provided as raw database dumps, participants utilized various preprocessing steps.

Both bookmarks and publications have an associated hash which identifies identical or more or less "similar" resources. However, the bookmark hash resembles the raw URL and different notations of the same webpage URL currently do not map to the same hash (e. g., `http://www.bibsonomy.org/` and `http://bibsonomy.org/` have different hashes). Thus, Tatu et al. [28] unified URLs to get a higher overlap. The same could be done with the publications, although BibSonomy's *simhash1* already includes some normalization (e. g., by removing non-numbers or letters from the title [44]). Further preprocessing included removal and extraction of additional data from the publication's *misc* field which sometimes contained additional keywords, conference names, or an English title, and removing LaTeX markup. Tatu et al. also filled missing BibTeX fields by accessing digital libraries like ACM, IEEE, or ScienceDirect. The most effort Tatu et al. put into the preprocessing of textual content. They normalized words to concepts using Wordnet, e. g., to map the variations *EU*, *European Union*, or *European Community* all to the concept *european_union*. Furthermore, they normalized tags to take care of spelling errors, synonyms, abbreviations, or joined concepts. By using Wordnet, this was only possible for English words, of course.

An interesting approach to reduce the impact of stop words from the title was adopted by M. Lipczak [25]. He scored the words by the frequency of their usage as tag. Additionally, he normalized words by removing non-numbers or letters and by making all letters lowercase – as suggested by our evaluation methodology (Section 3.3.1. In 2009, Lipczak et al. [26] further cleaned the data by removing posts that were imported from an external source into BibSonomy. They identified such posts by looking at sets of tags that were assigned to many posts by one user (e. g., the tag *indexforum* to 9,183 posts), or that have the same time stamp. They also separated the data into bookmark and publication posts.

3.4.2.3 Data Selection/Sources

All participants not only used the tags attached to posts in the training data as source to recommend tags, but also additional metadata contained in each post like title, description, or URL.

Using words from several columns of the database dumps and also tags as input for their multilabel text classification algorithm, Katakis et al. [20] decreased the dimensionality of their problem by taking only words and tags with a minimal frequency into account.

Lipczak et al. [25, 26] distinguished three sources to generate tags: resource title, resource tags, and personomy tags. This was motivated by the observation that there is a big variation of precision and recall. The title tags are extracted from the resource's title. Although those tags have a relatively low recall (less than 18 %), their precision is best. In contrast, the personomy tags, i. e., the tags used by the user at hand only, have an excellent recall of around 90 % but a precision of less than 0.1 %. This suggests that users often re-use tags. Lipczak also points out that high-frequency personomy tags likely represent topics or interests of the user (e. g., like *thesis* or *bookmarking*) while the large number of low-frequency tags might represent specific resources. Since only few resources have been tagged by two or more users, tags previously attached to a resource are not a good source. They also built tag-tag and title-tag co-occurrence graphs to exploit relations between tags as a source to gather "related tags".

Tatu et al. based their suggestions on normalized tags from posts and normalized concepts from textual content of resources. This includes user added text like title or description as well as the document content. Using NLP tools, they extracted important concepts from the textual metadata and normalize them using Wordnet.

Various external sources for tags were used by Mrosek et al. [31]: Delicious to directly get tags for URLs, Google Scholar[12] to gather web page and document links for publications, and finally the crawled content of those web pages and documents.

3.4.2.4 Personalization

Taking tags extracted from the resource title and from previous posts, M. Lipczak [25] uses the personomy tags to higher score those tags "which are most likely to be chosen by the user". This allowed him to better choose the lexical forms preferred by the user and to include user specific tags expressing certain interests. In the 2009 followup [26], Lipczak et al. additionally tried to capture different tagging situations, e. g., importing a batch of posts vs. posting a single publication (Section 3.4.2.5).

Interestingly, apart from that, only few participants applied personalization methods. Therefore, this could be a prolific field for future research (Section 3.7).

3.4.2.5 Ranking/Scoring

Most approaches relied on the frequency of a tag appearing in a particular source to calculate a score for the tag. The scores were often normalized to combine scores from different sources. E. g., Lipczak et al. [26] scored tags they extracted from the title (and URL for bookmarks) by their usage frequency in previous posts. They applied a threshold, below which they assigned a fixed score that estimates the usage-probabilty of low-frequency tags. A more sophisticated method was used for tags from the user's personomy. Each tag is scored by the number of distinctive days it was used in order to mitigate import and batch-tagging effects. Furthermore,

[12] `http://scholar.google.com/`

Lipczak et al merge scores of tags from several sources A and B using the formula for the union of independent probabilistic events $P(A \cup B) = P(A) + P(B) - P(A)P(B)$.

Due to the variety of different sources, Mrosek et al. [31] use several scoring strategies. For tags extracted from the content of web pages they consider how often the tag appears in different parts of the page (i. e., title, description, keywords, body). Similarly, for publication posts, they differentiate between the tag coming from the title, journal, or description field.

Ju and Hwang [21] applied a "limit condition" to eliminate keywords extracted from the resource metadata that are less often used as tags than the average. Their formula also penalizes frequent terms.

3.4.3 Results and Discussion

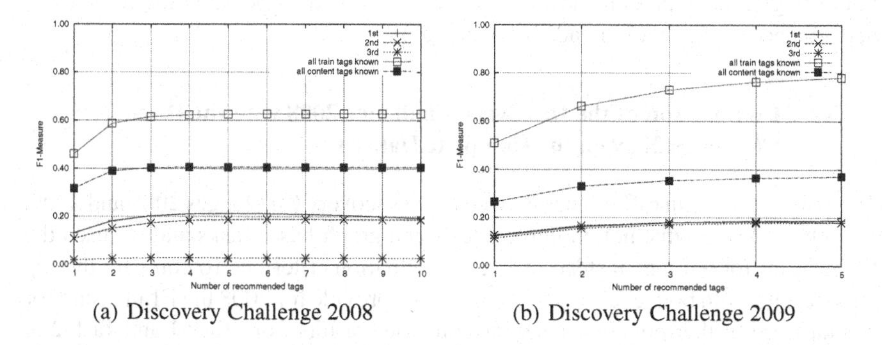

(a) Discovery Challenge 2008 (b) Discovery Challenge 2009

Fig. 3.2 The F1 measure of the top three approaches on the Task 1 datasets from 2008 and 2009. The winner was determined by the F1 measure for 10 tags in 2008 and for 5 tags in 2009.

3.4.3.1 Discovery Challenge 2008

The winners of the Discovery Challenge 2008 with an F1 measure of 0.193 for ten tags were M. Tatu and colleagues from Lymba Corporation, Richardson, Texas [28], see Figure 3.2(a). Close, with an F1 measure of 0.187 was M. Lipczak (Dalhousie University, Halifax, Canada) [25] as second. Katakis et al. [20] from Aristotle University of Thessaloniki came in third. As one can see in the plot, the F1 measure did not change much for more than five tags or was even decreasing. Tatu et al. had a constantly higher score, and only for ten tags Lipczak came close. The top two approaches have a score of approximately half of the best possible F1 measure for recommendations based on the content in the provided database dumps. The addition of further content, like full-text and Wordnet as done by Tatu et al., did not make a big difference.

3.4.3.2 Discovery Challenge 2009

In 2009, there were more participants and thus the competition was more tight. Lipczak et al. from Dalhousie University [26] were the winners of Task 1 with an F1 measure of 0.187 for five tags, see Table 3.2(b). The second of Task 1 (Mrosek et al. [31] from the University of Applied Sciences Gelsenkirchen), and the third (Ju and Hwang [21] from Soongsil University, Seoul) follow with an F1 measure of 0.180 each. As in 2008, the top participants had a score of about half of the content-based optimal recommender. The considerably better performance of the "all train tags known" baseline compared to the Discovery Challenge 2008 dataset can be explained by the larger amount of tags in the training data (Section 3.4.1). Interestingly, the approach shows clearly that it should be possible to increase the performance of the recommenders considerably.

3.4.3.3 Comparison of the Discovery Challenge 2008 with the Discovery Challenge 2009 on the Complete Dataset

There were two main differences between the Discovery Challenges 2008 and 2009: In 2008 we had no distinction into content- and graph-based tasks and we used the F1 measure for ten tags instead of five as evaluation criterion. To compare the top three results from both challenges, we therefore provide for 2008 the F1 measure for five tags and furthermore evaluated the combined results from Task 1 and Task 2 in 2009 (except for the second team that only participated in Task 1). Table 3.4 shows the results. First, one can see that there are no big differences between the f1m of the same approach in the different columns. In 2008, there is a small increase

Table 3.4 Comparison of the 2008 with the 2009 results. The bold values (f1m@10 in 2008 and f1m@5 in 2009) were used to determine the winner. The other two columns are for comparison purposes. The last column (f1m@5 combined) shows the results on the combined test data of Task 1 and Task 2 for the 2009 challenge for those teams that participated also in Task 2.

		2008		2009	
	participants	**f1m@10**	f1m@5	**f1m@5**	f1m@5 combined
2008	1st: Tatu et al. [28]	0.193	0.213		
	2nd: M. Lipczak [25]	0.187	0.188		
	3rd: Katakis et al. [20]	0.028	0.028		
2009	1st: Lipczak et al. [26]			0.187	0.190
	2nd: Mrosek et al. [31]			0.180	-
	3rd: Ju and Hwang [21]			0.180	0.182

when only five tags are regarded. Adding the data from Task 2 in 2009 only slightly increased the f1m since the 778 additional posts (Table 3.6) are few compared to the 43,002 posts of Task 1. The columns "f1m@5" for 2008 and "f1m@5 combined" for 2009 allow us to compare the performance of the top three approaches. Lipczak et al. only slightly improved their performance in 2009 but did not reach a better F1 measure than the first of 2008, Tatu et al. Overall, the performance of the top two (three for 2009) approaches is similar.

3.4.4 Lessons Learnt

Except for Katakis et al., the top participants did not apply sophisticated machine learning algorithms but rather simple methods. The main outcome of both challenges is that a clever combination of simple methods – e. g., as done by Lipczak et al. – yields good results. Moreover, they can be computed efficiently to produce online recommendations (Section 3.6). Overall, the participants tried to first gather a larger set of tag candidates from various sources (metadata of resources, full text of documents, crawled web pages, Wordnet) and then tried to rank them. This is probably the point where personalization becomes important – to select the tags from the candidates that best fit to the user's tagging behaviour. The "all train tags known" baseline indicates clearly that ongoing research has a lot of potential to increase the quality of the results.

3.5 Graph-Based Tag Recommendations

For Task 2 of the 2009 challenge, we ensured that every tag, user, and resource from the test dataset was already contained in the post-core at level 2 of the training data.

3.5.1 Datasets

In principle the participants could use the same *training data* as in Task 1 (Section 3.4.1) but we also provided a post-core at level 2 dataset (see Table 3.5). It has the property, that every resource, user, and tag occurs in at least two posts and is a subset of the training data of Task 1 [23]. The general format of the *test data* is as described in Section 3.3.2. Table 3.6 shows the number of items in the test datasets. All resources, tags, and users have the property of being present in the post-core at level 2, i. e., in the training data. The number of posts in the test data of this task (778) is considerably smaller than that of Task 1 in 2009 (43,002 posts, see Table 3.3).

Table 3.5 The number of posts, resources, tags, and users in the training data for Task 2.

	until Dec. 31st 2009		
	overall	bookmarks	publications
#posts	64,120	41,268	22,852
#resources	22,389	14,443	7,946
#users	1,185	861	788
#tags	13,252	-	-

Table 3.6 The number of posts, resources, tags, and users in the test data for Task 2.

	January 1st – June 30th 2009		
	overall	bookmarks	publications
#posts	778	431	347
#resources	667	387	280
#users	136	91	81
#tags	862	-	-

3.5.2 Approaches

Here we briefly describe the tag recommendation methods of the top results for Task 2. The structure of this section is different compared to Section 3.4.2 because some of the challenges of Task 1 (suggesting new tags, preprocessing, etc.) were not an issue in Task 2.

The winners, Rendle and Schmidt-Thieme [36], produced recommendations with a statistical method based on factor models. Therefore, they factorized the folksonomy structure to find latent interactions between users, resources and tags. Using a variant of the stochastic gradient descent algorithm, the authors optimized an adaptation of the Bayesian Personal Ranking criterion [35]. The learned factor models are ensembled using the rank estimates to remove variance from the ranking estimates. Finally, the authors estimated how many tags to recommend to further improve precision using a linear combination of three estimates. Balby Marinho et al. [29] used relational classification methods in a semi-supervised learning scenario to recommend tags. They built a weighted graph with user-resource tuples as nodes that are connected when they contain the same user or resource. The weights are based on different weighting schemes that incorporate user-tag, user-resource, and resource-tag profiles. For a test post, they recommended tags from related posts in its neighborhood using a probabilistic weighted average. Lipczak et al. [26] were also the third of Task 2, for a description of their method (Section 3.4.2).

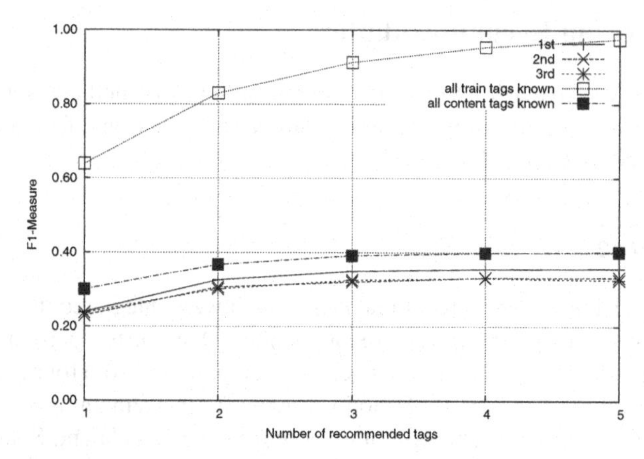

Fig. 3.3 The F1 measure for the Task 2 of the 2009 challenge.

3.5.3 *Results and Discussion*

The winners of Task 2, Rendle and Schmidt-Thieme from the University of Hildesheim, Germany [36] achieved an F1 measure of 0.356. The second, Balby Marinho et al. [29] – also from the University of Hildesheim, Germany – achieved an F1 measure of 0.332, followed by the winning recommender from Task 1, Lipczak et al. [26], with an F1 measure of 0.325. The "all train tags known" baseline is the upper limit. It does not reach an F1 measure of 1 because some posts have more than five tags. Although all participants are far away from this theoretical boundary, the winner almost reaches the "all content tags known" baseline. Compared to Task 1 (Figure 3.2(b)) the performance of the top three recommenders is higher.

3.5.4 *Lessons Learnt*

First, one can say that knowing some tags the user has used before or that have been attached to the resource before considerably improves the quality of tag recommendations. An F1 measure that is almost twice as high as on the "unknown" data of Task 1 shows this. Second, sophisticated machine learning approaches like the ones by Rendle and Schmidt-Thieme [36] or Balby Marinho et al. [29] are able to produce good recommendations on such data. Nevertheless, simpler methods like the one from Lipczak et al. [26] are competitive and often easier to compute. Third, only a small percentage of posts in a collaborative tagging system is from known users with known resources. Because of this, Task 1 is the more realistic one. Another lesson of the Discovery Challenges 2008 and 2009 is that methods that cleverly combine simple tag recommendations from various sources have an acceptable – and even high – performance.

3.6 Online Tag Recommendations

In this section we give a brief overview on the setting, the methods some recommenders used and the resulting recommendation performances of the online task (Task 3) of the Discovery Challenge 2009.

3.6.1 *Setting*

The participants implemented recommenders which were integrated into the framework using instances of the remote recommender. Overall, ten participants from seven countries deployed 13 recommenders – seven of them (from four participants) were running on machines in BibSonomy's network, the remaining six were distributed all over the world (amongst other countries, in Canada and South Korea). All recommenders had to respect a timeout of 1000 ms between the sending of the recommendation request and the arrival of the result. If they failed to deliver their recommendation in time, we set precision and recall for the corresponding post to zero, and showed to the user another recommendation.

The recommendations were evaluated over the period from July 29th to September 2nd 2009. During that time, more than 28,000 posting processes had to be handled, where each recommender was randomly selected to deliver recommendations for at least 2,000 posts. For evaluation, we regarded only public, non-spam posts and therefore the results in the next section are based on approximately 380 relevant posting processes per recommender (for the exact counts, see Table 3.7).

Table 3.7 The number of posts regarded for evaluation.

recommender-id	3	5	6	7	12	13	14	16
#posts	347	391	361	415	385	380	370	398

Although 13 recommenders participated in the online task, only eight of them managed to deliver results in at least 50 % of all requested posting processes. The remaining five recommenders answered only in less than 5 % of all cases and are thus ignored in Table 3.7 and in the figures and discussion following in Section 3.6.3.

3.6.2 *Approaches*

The winning recommender 6 is the approach by Lipczak et al. [26] who were also successful in Task 1 (1st) and Task 2 (3rd). For a description of their method, we refer to Section 3.4.2. Recommender 3 by Si et al. [39] performed so called "Feature Driven Tagging" by extracting and weighting features like words, ids, hashes, phrases from the resources. Each feature then generates a list of tags. The weight of the features is estimated using TF×IDF and TF×ITF (term frequency × inverted document frequency and term frequency × inverted tag frequency [1]); the tags of the features are determined using co-occurrence counts, mutual information, and χ^2 statistics. For recommender 5, Cao et al. [8] divided the posts in four categories, depending on the case if the user or resource of the post is known or not. Then, for each category they learned a model to rank the tags using a ranking SVM. To augment the available tags for posts (besides the full text and the tags of the resource), the authors used post-content similarity and k-Nearest-Neighbors.

3.6.3 Results and Discussion

First, we have a look at precision and recall of each recommender in the evaluation mode relevant for the challenge (Figure 4(a)). For a posting process in which the recommender could not deliver a recommendation in time, precision and recall were set to zero. In this setting, recommender 6 by Lipczak et al [26] is clearly the winner with an F1 measure of 0.205 for five tags. This is slightly better than the recommender's performance in the offline Task 1 (Figure 3.2(b)) where it had an F1 measure of 0.187. The performance of the remaining methods varies between an F1 measure of 0.030 and 0.171 for five tags – all those recommenders have a recall of less than 0.2.

If we disregard the timeout limit of 1000 ms and also evaluate the suggestions which came later (Figure 4(b)), we get a different picture. Of course, all recommenders improve – but in particular recommender 14 gains both precision and recall. This can be explained by the latency of recommender 14. Although it returned a suggestion in almost as many posts as the winning recommender 6, only 20 % of the posts were delivered in time – in contrast to almost 80 % of the posts for recommender 6.

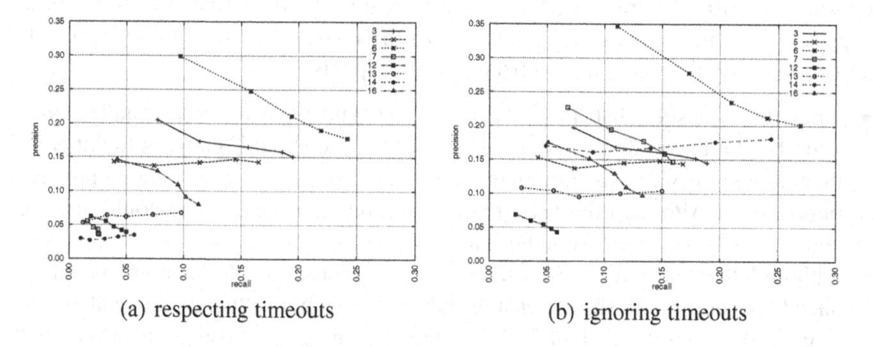

(a) respecting timeouts (b) ignoring timeouts

Fig. 3.4 Recall and precision of the deployed recommenders.

3.6.4 Lessons Learnt

The comparison and evaluation of different recommendation approaches is necessary to demonstrate the progress made by newly developed methods. Usually, a real world dataset which contains the predicted information is used to demonstrate the power of these methods. Such a subsequent analysis is a good way to do the evaluation but has its limitations, since recommendation applications running in a real world system need to fulfill further requirements. Besides having a good recommendation quality, the system has to be fast and robust in such a setting. This was one important lesson learnt from the online task.

More generally it turns out that any kind of competition is an important instrument to foster new research directions. The two challenges we have organized have attracted many researchers. The Netflix prize yielded a boost on a larger scale. Such challenges have stimulated other companies to open their data for researchers. One example is RichRelevance Inc. which has started in 2010 a project called RecLab Core[13]. Basically, they have defined a recommendation API which allows researchers to implement their own recommender. Both steps, the model learning and the model application step are supported. In the end, the model can be used to compute recommendations in an online setting for their customers.

3.7 Summary and Future Challenges

The results of the ECML PKDD Discovery Challenges 2008 and 2009 have shown that clever combinations of simple methods yield tag recommendation results which can outperform state-of-the-art machine learning approaches. If more sophisticated methods will gain performance is thus at least questionable, in particular since they are often not able to solve the cold-start problem, i. e., to deliver recommendations for unknown users or resources.

Before we focus on future challenges in tag recommendation systems, we bring up some proposals for improving the evaluation of tag recommender systems. These suggestions stem from our experience with organizing the two Discovery Challenges and from feedback we got from the participants.

- One should ensure that all posts in the test data were really manually annotated by the users of the system. This means, one should ignore automatically tagged posts, e. g., posts imported from other systems, when evaluating tag recommenders. Although the tags might be human-generated, they could also be automatically extracted from the title.
- Although the distribution of tags, users, and resources in a folksonomy is unbalanced by nature, one should ensure that this does not influence the evaluation. E. g., in the 2009 test data of Task 1, there was one user whose publication posts attributed a large amount of the overall posts (approx. 65 %). Similarly, 9,183 of the 16,194 bookmark posts (ap. 56 %) of the 2009 test dataset were posted by one user which always used the same tag ("indexforum").

[13] `http://code.richrelevance.com/reclab-core/`

- The most important suggestion is to focus on online challenges, since they better resemble a real recommendation task. They are, however, much more difficult to handle for both organizers and participants.

Since the emergence of collaborative tagging systems, research in the field of tag recommendations has established evaluation protocols, defined baselines, and came up with quite some valuable methods. Here we present a selection of challenges which could be promising entry points for the next generation of tag recommendation algorithms.

Different Types of Tags. Most of the approaches that tackle the problem of tag recommendations in folksonomies are rather generic (e. g., based on Collaborative Filtering, co-occurrence counts, or the content of resources) and do not distinguish different types of tags. However, Golder and Huberman [14] identified seven kinds of functions which tags can perform for a resource: identifying what (or whom) it is about, identifying what it is, identifying who owns it, refining categories, identifying qualities or characteristics (*scary*, *funny*, *stupid*, *inspirational*), self reference (*myown*), and task organizing (*toread*, *thesis*). Naturally, a first step to further improve tag recommendations would require methods to classify tags into one of those categories. Steps into this direction have been done by Strohmaier [42] who tries to identify *purpose tags*, i. e., tags that describe the intent rather than the content. Once different categories can be distinguished, one could then focus the recommender on tags from categories where it is easier to 'guess' the tags for a resource. E. g., it might be rather difficult to recommend tags which describe the organization of tasks like *toread*, since it requires to know why the user is tagging the resource at hand.

Personalization. An important challenge is the personalization of recommendations by respecting the different tagging habits of users. A tag, for example, can be in different categories, depending on the user's understanding. Since tagging allows the user to capture his view on a certain resource (besides merely describing its content), methods to predict the opinion of a user regarding a resource could lead to more personalized recommendations. Therefore, methods are necessary which identify tags expressing opinions about items. Most existing approaches try to identify the opinion of users based on written texts.[14] Breck et al. [6], for example, suggest a method to identify words and phrases in texts which express opinions. Recommending such 'opinion tags' without influencing the user by pushing his opinion into a certain direction is a challenging task – it has been shown by Cosley et al. [10] that recommender systems affect the user's opinion.

Another option could be the integration of semantic approaches to capture the user's personal information model [38]. This could improve personalization, since tags could then explicitly be assigned to a category or distinguished from other tags by the user.

In general, however, too personalized tag recommendations hamper the goal of consolidating the tag vocabulary across users. A recommender which perfectly adapts to the language and thinking of the user does not give him the chance to see

[14] For a survey on opinion mining and sentiment analysis see [33].

other users' vocabulary and thereby adopt popular tags or follow accepted tagging habits. Finding the balance between recommending what the user wants and what is good for him, the system, or other users is a difficult task and a widely discussed topic in the recommender community.

Trust. A relatively new topic in the recommender community is the incorporation of *trust between users* into the systems. This can be seen as a special way of personalization. Although it is very difficult to define trust (personal background, history of interaction, etc. play a role), in the context of recommender systems, trust usually implies the similarity of users in their opinion about a topic. It is therefore important to take into account the context based on which trust has been computed – someone a user would trust in recommending a movie might not be trustworthy in providing a recommendation for a certain product. Recommendation methods like Collaborative Filtering, which incorporate similarity between users, can use trust instead of or in combination with similarities. Trust can also be used to filter or sort information.

Up to now, there is no attempt known to leverage trust for tag recommendations – some typical applications are in the movie domain (see the survey by Golbeck [13]). One reason for that might be the lack of appropriate data to compute trust in folksonomies (like ratings). Although most systems provide some kind of social networking features (groups, friends), this data often is not easily available to the research community and it is also questionable if it is a useful basis to compute trust in tagging. Both friends and group members might have an interest in the same topics the user has, however, their tagging behaviour or used vocabulary might be quite different.

Another category of trust is the *user's trust in the recommendation system* or *impersonal trust* [32]. It describes to what respect the user trusts the system to give recommendations which are helpful to him. One approach to strengthen the user's trust is the *explanation* of recommendations [16] which makes recommendations more traceable – a well-known example are Amazon's "Customers Who Bought This Item Also Bought" recommendations. Vig et al. [43] find that tags can be used to explain movie recommendations and thereby help the user to understand why an item was recommended and to decide if they like it. Although up to now there is no research known on explaining tag recommendations, depending on the used algorithm existing approaches can be adopted, e. g., [16] for Collaborative Filtering. One hindrance in building up trust is spam, since it can influence recommendation systems and degrade the quality of recommendations [24].

Further Aspects. Systems like BibSonomy, which allow users to maintain a relation between tags suggest that more complex forms of recommendations might be necessary. I. e., instead of recommending just tags, one could also provide more structure by recommending elements for the user's tag relation. Beyond tagging, one could incorporate ontology learning techniques [7] and discuss the recommendation of concepts, general relations or even (parts of) ontologies. This has been partly adressed by Haase et al [15], who use a Collaborative Filtering approach to suggest personalized ontology changes.

Acknowledgements. Part of this research was funded by the European Union in the Nepomuk (FP6-027705) and Tagora (FET-IST-034721) projects.

References

1. Baeza-Yates, R.A., Ribeiro-Neto, B.: Modern Information Retrieval. Addison-Wesley Longman Publishing Co., Inc., Boston (1999)
2. Bennett, J., Lanning, S.: The netflix prize. In: Proceedings of the KDD Cup Workshop 2007, pp. 3–6. ACM, New York (2007)
3. Bennett, J., Elkan, C., Liu, B., Smyth, P., Tikk, D.: Kdd cup and workshop 2007. SIGKDD Explorations Newsletter 9(2), 51–52 (2007)
4. Benz, D., Hotho, A., Jäschke, R., Krause, B., Mitzlaff, F., Schmitz, C., Stumme, G.: The social bookmark and publication management system BibSonomy. The VLDB Journal 19(6), 849–875 (2010)
5. Bogers, T.: Recommender Systems for Social Bookmarking. PhD thesis, Tilburg University, Tilburg, The Netherlands (December 2009)
6. Breck, E., Choi, Y., Cardie, C.: Identifying expressions of opinion in context. In: IJCAI 2007: Proceedings of the 20th International Joint Conference on Artifical Intelligence, pp. 2683–2688. Morgan Kaufmann Publishers Inc., San Francisco (2007)
7. Buitelaar, P., Cimiano, P., Magnini, B. (eds.): Ontology Learning from Text: Methods, Evaluation and Applications. Frontiers in Artificial Intelligence, vol. 123. IOS Press (July 2005)
8. Cao, H., Xie, M., Xue, L., Liu, C., Teng, F., Huang, Y.: Social tag prediction base on supervised ranking model. In: Eisterlehner, et al. (eds.) [11], pp. 35–48
9. Cattuto, C., Loreto, V., Pietronero, L.: Collaborative tagging and semiotic dynamics. CoRR, abs/cs/0605015 (May 2006)
10. Cosley, D., Lam, S.K., Albert, I., Konstan, J.A., Riedl, J.: Is seeing believing?: how recommender system interfaces affect users' opinions. In: CHI 2003: Proceedings of the SIGCHI Conference on Human Factors in Computing Systems, pp. 585–592. ACM, New York (2003)
11. Eisterlehner, F., Hotho, A., Jäschke, R. (eds.): ECML PKDD Discovery Challenge (DC 2009), vol. 497 (September 2009), CEUR-WS.org
12. Fellbaum, C. (ed.): WordNet An Electronic Lexical Database. The MIT Press, Cambridge (1998)
13. Golbeck, J.: Trust on the world wide web: A survey. Foundations and Trends in Web Science 1(2), 131–197 (2006)
14. Golder, S., Huberman, B.A.: The structure of collaborative tagging systems. Journal of Information Science 32(2), 198–208 (2006)
15. Haase, P., Hotho, A., Schmidt-Thieme, L., Sure, Y.: Collaborative and Usage-Driven Evolution of Personal Ontologies. In: Gómez-Pérez, A., Euzenat, J. (eds.) ESWC 2005. LNCS, vol. 3532, pp. 486–499. Springer, Heidelberg (2005)
16. Herlocker, J.L., Konstan, J.A., Riedl, J.: Explaining collaborative filtering recommendations. In: CSCW 2000: Proceedings of the 2000 ACM Conference on Computer Supported Cooperative Work, pp. 241–250. ACM, New York (2000)
17. Herlocker, J.L., Konstan, J.A., Terveen, L.G., Riedl, J.T.: Evaluating collaborative filtering recommender systems. ACM Trans. Inf. Syst. 22(1), 5–53 (2004)
18. Hotho, A., Jäschke, R., Schmitz, C., Stumme, G.: Information Retrieval in Folksonomies: Search and Ranking. In: Sure, Y., Domingue, J. (eds.) ESWC 2006. LNCS, vol. 4011, pp. 411–426. Springer, Heidelberg (2006)

19. Hotho, A., Krause, B., Benz, D., Jäschke, R. (eds.): ECML PKDD Discovery Challenge (RSDC 2008) (2008)
20. Tsoumakas, G., Katakis, I., Vlahavas, I.: Multilabel text classification for automated tag suggestion. In: Hotho, et al. (eds.) [19], pp. 75–83
21. Ju, S., Hwang, K.-B.: A weighting scheme for tag recommendation in social bookmarking systems. In: Eisterlehner, et al. (eds.) [11], pp. 109–118
22. Jäschke, R.: Formal Concept Analysis and Tag Recommendations in Collaborative Tagging Systems. Dissertationen zur Künstlichen Intelligenz, vol. 332. Akademische Verlagsgesellschaft AKA, Heidelberg, Germany (January 2011)
23. Jäschke, R., Marinho, L., Hotho, A., Schmidt-Thieme, L., Stumme, G.: Tag recommendations in social bookmarking systems. AI Communications 21(4), 231–247 (2008)
24. Lam, S.K., Riedl, J.: Shilling recommender systems for fun and profit. In: WWW 2004: Proceedings of the 13th International Conference on World Wide Web, pp. 393–402. ACM, New York (2004)
25. Lipczak, M.: Tag recommendation for folksonomies oriented towards individual users. In: Hotho, et al. (eds.) [19], pp. 84–95
26. Lipczak, M., Hu, Y., Kollet, Y., Milios, E.: Tag sources for recommendation in collaborative tagging systems. In: Eisterlehner, et al. (eds.) [11], pp. 157–172
27. Lipczak, M., Milios, E.: Learning in efficient tag recommendation. In: Proceedings of the Fourth ACM Conference on Recommender Systems, RecSys 2010, pp. 167–174. ACM, New York (2010)
28. Srikanth, M., Tatu, M., D'Silva, T.: Rsdc 2008: Tag recommendations using bookmark content. In: Hotho, et al. (eds.) [19], pp. 96–107
29. Marinho, L.B., Preisach, C., Schmidt-Thieme, L.: Relational classification for personalized tag recommendation. In: Eisterlehner, et al. (eds.) [11], pp. 7–16
30. Milicevic, A., Nanopoulos, A., Ivanovic, M.: Social tagging in recommender systems: a survey of the state-of-the-art and possible extensions. Artificial Intelligence Review 33(3), 187–209 (2010)
31. Mrosek, J., Bussmann, S., Albers, H., Posdziech, K., Hengefeld, B., Opperman, N., Robert, S., Spira, G.: Content- and graph-based tag recommendation: Two variations. In: Eisterlehner, et al. (eds.) [11], pp. 189–200
32. O'Donovan, J., Smyth, B.: Trust in recommender systems. In: IUI 2005: Proceedings of the 10th International Conference on Intelligent User Interfaces, pp. 167–174. ACM, New York (2005)
33. Pang, B., Lee, L.: Opinion mining and sentiment analysis. Foundations and Trends in Information Retrieval 2(1-2), 1–135 (2008)
34. Rae, A., Sigurbjörnsson, B., van Zwol, R.: Improving tag recommendation using social networks. In: Adaptivity, Personalization and Fusion of Heterogeneous Information, RIAO 2010, pp. 92–99. Le Centre De Hautes Etudes Internationales d'Informatique Documentaire, Paris (2010)
35. Rendle, S., Freudenthaler, C., Gantner, Z., Lars, S.-T.: Bpr: Bayesian personalized ranking from implicit feedback. In: Proceedings of the Twenty-Fifth Conference on Uncertainty in Artificial Intelligence, UAI 2009, pp. 452–461. AUAI Press, Arlington (2009)
36. Rendle, S., Schmidt-Thieme, L.: Factor models for tag recommendation in BibSonomy. In: Eisterlehner, et al. (eds.) [11], pp. 235–242
37. Said, A., Berkovsky, S., De Luca, E.W.: Putting things in context: Challenge on context-aware movie recommendation. In: Proceedings of the Workshop on Context-Aware Movie Recommendation, pp. 2–6. ACM, New York (2010)
38. Sauermann, L.: The gnowsis – using semantic web technologies to build a semantic desktop. Diploma thesis, Technical University of Vienna (2003)

39. Si, X., Liu, Z., Li, P., Jiang, Q., Sun, M.: Content-based and graph-based tag suggestion. In: Eisterlehner, et al. (eds.) [11], pp. 243–260
40. Song, Y., Zhang, L., Lee Giles, C.: Automatic tag recommendation algorithms for social recommender systems. Transactions on the Web 5(1), 1–31 (2011)
41. Sood, S., Owsley, S., Hammond, K., Birnbaum, L.: TagAssist: Automatic Tag Suggestion for Blog Posts. In: Proceedings of the International Conference on Weblogs and Social Media, ICWSM 2007 (2007)
42. Strohmaier, M.: Purpose tagging: capturing user intent to assist goal-oriented social search. In: SSM 2008: Proceedings of the 2008 ACM Workshop on Search in Social Media, pp. 35–42. ACM, New York (2008)
43. Vig, J., Sen, S., Riedl, J.: Tagsplanations: explaining recommendations using tags. In: IUI 2009: Proceedings of the 13th International Conference on Intelligent User Interfaces, pp. 47–56. ACM, New York (2009)
44. Voss, J., Andreas, H., Robert, J.: Mapping bibliographic records with bibliographic hash keys. In: Kuhlen, R. (ed.) Proceedings of the ISI Information: Droge, Ware oder Commons? Hochschulverband Informationswissenschaft. Verlag Werner, Hülsbusch (2009)

Chapter 4
A Multi-criteria Approach for Automatic Ontology Recommendation Using Collective Knowledge

Marcos Martínez-Romero, José M. Vázquez-Naya, Javier Pereira, and Alejandro Pazos

Abstract. Nowadays, ontologies are considered an important tool for knowledge structuring and reusing, especially in domains in which the proper organization and processing of information are critical issues (e.g. biomedicine). In these domains, the number of available ontologies has grown rapidly during the last years. This is very positive because it enables a more effective (or more intelligent) knowledge management. However, it raises a new problem: what ontology should be used for a given task? In this work, an approach for the automatic recommendation of ontologies is proposed. This approach is based on measuring the adequacy of an ontology to a given context according to three independent criteria: (i) the extent to which the ontology covers the context, (ii) the semantic richness of the ontology in the context, and (iii) the popularity of the ontology in the Web 2.0. Results show the importance of using collective knowledge in the fields of ontology evaluation and recommendation.

4.1 Introduction

In recent years, ontologies have become an essential tool to structure and reuse the increasing growth of information in the Web. Ontologies make it possible to describe domain entities and their relationships in a formal and explicit manner, facilitating the exchange of information between people or machines that use different representations to refer to the same meaning. In addition, knowledge represented

Marcos Martínez Romero · José M. Vázquez Naya · Javier Pereira
IMEDIR Center, University of A Coruña, 15071 A Coruña, Spain
e-mail: {marcosmartinez,jmvazquez,javierp}@udc.es

Alejandro Pazos
Department of Information and Communication Technologies, Computer Science Faculty, University of A Coruña, 15071 A Coruña, Spain
e-mail: apazos@udc.es

J.J. Pazos Arias et al.: Recommender Systems for the Social Web, ISRL 32, pp. 89–103.
springerlink.com

using ontologies can be processed by computers, making it possible to interpret and manage large amounts of information automatic and semantically.

Reusing existing ontologies rather than creating new ones is a desirable practice. Building an ontology from scratch is a very complex and time-consuming task that requires specialized human resources. In addition, avoiding the existence of multiple ontologies that represent the same knowledge is necessary to ensure proper interoperability. However, due to the increasing number, complexity and variety of existing ontologies, frequently containing thousands of concepts and relations between them, choosing the ontology or ontologies to be reused in a semantic annotation problem or to design a specific application is a difficult task. Due to this, the development of approaches and tools that facilitate the task of selecting the best ontology or ontologies for a given context and task is becoming a priority for researchers.

Many criteria may be taken into account when recommending an ontology for a given task: the number of concepts and relationships that it contains, the location of concepts in the ontology, the language in which it is represented, etc. However, it is necessary to select the most relevant criteria and combine them accurately, in order to make it possible to obtain good results in a reasonable time and without the intervention of an expert.

This paper presents an approach for ontology recommendation that combines the assessment of three criteria: context coverage, semantic richness and popularity. The combination of these different, but complimentary evaluation methods is aimed to give a robust reliability and performance in large-scale, heterogeneous and open environments like the Semantic Web.

4.2 Related Work

Ontology recommendation (also known as ontology selection) is the process that allows identifying one or more ontologies that satisfy certain criteria [1]. This process is, in essence, an ontology evaluation problem. The need for evaluation strategies in the field of ontologies showed up as soon as 1994 with the work of Gómez Pérez in Stanford [2, 3]. However, since the Semantic Web was conceived in 2001 by Berners-Lee [4], the interest in this issue has increasingly grown. Since that moment, several works proposed to evaluate the quality of an ontology on the basis of ontology metrics such as the number of classes, properties, taxonomical levels, etc. [5]. More recently, during the last five years, different strategies have been proposed to select the best ontology for a particular context or task [6, 7, 8, 9, 10]. However, in general, these approaches have several drawbacks. Some of them are the following: (i) most approaches are not completely automatic and require an expert who guides the recommendation process; (ii) the input is restricted to a single keyword; (iii) the popularity of ontologies is not taken into account, or it is not correctly assessed [11]; (iv) ontologies are frequently considered simple sets of terms, ignoring the semantics provided by the relations between concepts.

In this paper, a new approach for ontology recommendation is proposed, which intends to overcome these drawbacks by means of the adequate combination of three different evaluation criteria. To the best of our knowledge, this approach is the first that aims to recommend the most appropriate ontologies for a given context using collective knowledge extracted from the Web 2.0, and also taking into account the inner richness of the ontologies and the context coverage provided by them. According to the complete review of ontology evaluation strategies published by Brank et al. [12], our approach fits into the category of multiple-criteria approaches.

4.3 An approach for Automatic Ontology Recommendation

In this section, the approach for automatic ontology recommendation is presented and explained. On the basis of the overall workflow shown in Figure 4.1, we start describing how the initial terms are semantically cleaned and expanded. Then, we illustrate the evaluation methods that constitute the core of the recommendation strategy. Finally, we describe the method followed to aggregate the different obtained scores into a unique score.

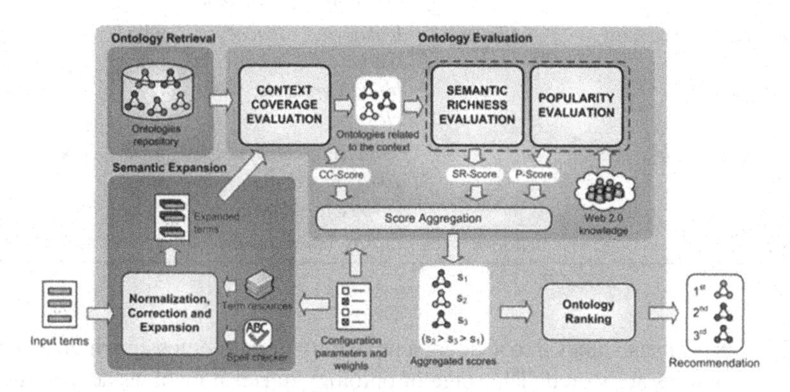

Fig. 4.1 Workflow of the ontology recommendation process

4.3.1 Semantic Expansion and Ontology Retrieval

The process starts from a set of initial terms that represent a given context. These terms may be provided by a user, a Semantic Web agent, or automatically identified from an information resource (e.g. text, image, webpage, etc.). The purpose of the semantic expansion stage is to expand each initial term with other terms with the same meaning (synonyms). These synonyms will be useful later, during the ontology evaluation process, to assess to what extent the initial terms are covered by each candidate ontology.

Firstly, the initial terms are **normalized** to a standard format. All up case letters are converted to down case. Punctuation, double spaces and special symbols are removed. Secondly, a spell checker is used to **correct** possible typographical mistakes. After that, one (or several) terminological resources (e.g. WordNet[1] as the general-purpose lexical resource for the English language, or UMLS[2] as a specialized resource for the biomedical field), are used to remove possible initial terms with the same meaning and, finally, each remaining term is **expanded** with its synonyms to increase the possibility of finding it into the ontologies.

Example 1. Suppose that the initial set of terms is *{KNEE, Leucocite., White_Cell, torax}* and that the selected terminological resource is WordNet. Then, the normalized, corrected and expanded terms are shown in Table 4.1. The term *white cell* would be removed, because it is a synonym of *leukocyte*.

Table 4.1 Normalization, correction and expansion. A "-" means that there were no changes

Term	Normalization	Correction	Expansion	Removed?
KNEE	*knee*	-	*knee, knee joint, human knee, articulatio genus, genu*	No
Leucocite	*leucocite*	*leukocyte*	*leukocyte, leucocyte, white blood cell, white cell, white blood corpuscle, white corpuscle, wbc*	No
White_Cell	*white cell*	-	*white cell, leukocyte, leucocyte, white blood cell, white blood corpuscle, white corpuscle, wbc*	Yes
torax	-	*thorax*	*thorax, chest, pectus*	No

The approach also takes into account a repository of ontologies, in which all candidate ontologies are stored. The stage of ontology retrieval involves accessing the repository and obtaining all the candidate ontologies, in order to start the evaluation process, which is described in the following sections.

4.3.2 Ontology Evaluation

The core of the ontology recommendation process is ontology evaluation. The presented approach proposes to assess each ontology according to three types of evaluation: (i) evaluation of context coverage, which consists on measuring the amount of input terms whose meaning is contained in the ontology; (ii) evaluation of semantic richness, or assessing the level of detail that the ontology provides; and (iii) evaluation of popularity, or relevance of the ontology in the community.

[1] http://wordnet.princeton.edu
[2] http://www.mln.nih.gov/research/umls

Each kind of evaluation produces a numerical score. The three scores obtained for each ontology are aggregated into a single value, which indicates the goodness of the ontology to describe the initial terms. In the following, these three evaluation strategies are explained. After that, we will explain how the score aggregation process is performed.

4.3.2.1 Context Coverage Evaluation

This step is addressed to obtain the ontologies that (partially or completely) represent the context. This is achieved by evaluating how well each ontology from a given repository covers the initial terms. We propose a metric, called *CCscore*, which consists on counting the number of initial terms that are contained in the ontology. An initial term is contained in the ontology if there is a concept in the ontology whose name matches the term or one of its synonyms. This metric only has into account exact matches, because it has been found that considering partial matches (e.g. *anaerobic, aerosol, aeroplane*) may reduce the search quality [13]. The exact matching function is defined as follows.

Definition 1 (*exact matching*). Let T be a set of terms and t a term from T. O is a set of ontologies, and o is an ontology from O. Let $expand(t)$ be the function that provides the synonyms for a term t. Also let $conceptnames(o)$ be the function whose output are the names for all the concepts in the ontology o. Then:

$$exactmatching(t,o) = \begin{cases} 1 & \text{if } expand(t) \cap conceptname(o) \neq \varnothing \\ 0 & \text{otherwise} \end{cases}$$

And the *CCscore* is defined as follows.

Definition 2 (*CCscore*). Let $T = \{t_1, t_2, \ldots, t_n\}$ be the set of initial terms after the steps of normalization and correction, with $n = |T|$. The *Context Coverage Score (CCscore)* for an ontology o is defined in the following manner:

$$CCscore(o,T) = \frac{\sum_{i=1}^{n} exactmatching(t_i, o)}{n}$$

Example 2. Suppose that the recommendation process starts from a set of terms $T = \{leukocyte, blood, neurone\}$ such that: $expand(leukocyte) = \{leukocyte, leucocyte, white\ blood\ cell, white\ cell\}$; $expand(blood) = \{blood\}$; and $expand(neurone) = \{neurone, neuron, nerve\ cell\}$. Also, suppose that we are working with an ontology o, which contains only four concepts, such that $conceptnames(o) = \{cell, white\ cell, neuron, tooth\ cell\}$. Then $exactmatching(leukocyte, o) = 1$; $exactmatching(blood, o) = 0$; and $exactmatching(neurone, o) = 1$. According to this, the function *CCscore* would be calculated as: $CCscore(o,T) = 1/3 + 0/3 + 1/3 = 2/3 = 0{,}67$.

4.3.2.2 Semantic Richness Evaluation

Ontologies with a richer set of elements about a specific context are potentially more useful to describe such context than more simple ones. In this section, we present a metric to assess the semantic richness of an ontology in a given context, which is called *Semantic Richness Score (SRscore)*.

Example 3. Suppose that it is necessary to select the best ontology to represent the term *neuron*. If we consider the ontologies *A* and *B* from Figure 4.2, it is possible to see that both cover that term. However, it is clear that the ontology *B* provides more detailed knowledge about that term than the ontology *A*. Ontology *B* has a definition of *neuron*, several synonyms, a variety of concepts related through the *part_of* relation, and diverse subclasses. If the term *neuron* is described with the ontology *B*, this description will be potentially more useful, both for users who want to access the ontology to retrieve information about the concept *neuron*, and for software agents that require to process the information from that concept in order to achieve inferences, or other operations. This basic example shows the importance of considering the semantic richness during the evaluation process.

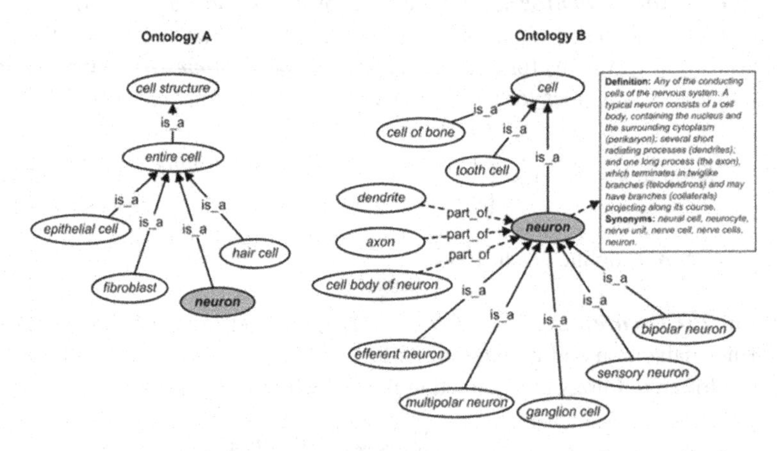

Fig. 4.2 Two ontologies that provide different levels of semantic richness for the concept *neuron*

The *SRscore* metric proposed in this paper is an improvement over the semantic richness metric presented in our previous work [14]. It is independent of the language in which the ontology is expressed (e.g. OWL) and it only evaluates the richness of the concepts that are involved in the context coverage, instead of taking into account the semantic richness of all the concepts in the ontology.

The *SRscore* for each ontology is calculated on the basis of the results obtained after three evaluation steps: (i) evaluation of concept relatives; (ii) evaluation of additional information; and (iii) evaluation of similar knowledge. In the following, these aspects are explained.

Evaluation of Concept Relatives

The amount of relatives for each concept is one of the aspects that may be considered important when measuring the semantic richness of an ontology. The evaluation of concept relatives is addressed to measure the average of relatives for the concepts in an ontology that cover the input terms. This is expressed by using an index called *relatives index*, which is defined as follows.

Definition 3 (*relatives index*). Suppose that o is an ontology whose set of concepts is C, and $c \in C$ that is a concept from o. Let P, S and H be the sets of concepts from o that are direct parents, siblings and children of c, respectively. Then, the number of relatives for a given concept c is expressed as:

$$relatives(c) = |P| + |S| + |H|$$

Suppose that $D = \{d_1, d_2, \ldots, d_n\} \subseteq C$ is the subset of concepts in o that cover the input terms. Then, the *relatives index* (r_i) for D would be calculated as:

$$ri(D) = \frac{\sum_{i=1}^{n} relatives(d_i)}{n}$$

Example 4. Suppose that D is a subset of the concepts in the ontology B (see Figure 4.2), such that $D = \{cell, axon, neuron\}$. Then:

$$ri(D) = \frac{relatives(cell) + relatives(axon) + relatives(neuron)}{3} = \frac{3+0+8}{3} = 3.67$$

Evaluation of Additional Information

Apart from its relatives, an ontology typically contains other information for each concept (e.g. relations with other concepts, definitions, etc.). The evaluation of additional information is addressed to measure this kind of information through a numerical index called *additional information index*, which is described in the following.

Definition 4 (*additional information index*). Suppose that C is the set of concepts of an ontology. We consider that the additional information (function *additionalinf*) is the number of characteristics of a concept $c \in C$ that provide information about such concept, with the exception of: (i) its relatives, which were already assessed; (ii) the concept name (or label), because we assume that every concept will have a name, and we will not consider it a distinguishing feature; and (iii) the concept instances, because we believe that they do not necessarily reflect the richness of the conceptual structure itself. Having this into account, the additional information for a concept would consist on the number of relations with other concepts, definitions, synonyms, restrictions over values or datatypes, etc.

Example 5. The concept *neuron* in the ontology B (see Figure 4.2) has 1 definition, 6 synonyms and 3 concepts related through the *part_of* relation. In this case, $additionalinf(neuron) = 1+6+3 = 10$.

Suppose that o is an ontology whose set of concepts is C. Also suppose that $D = \{d_1, d_2, \ldots, d_n\} \subseteq C$ is the subset of concepts in o that cover the input terms. Then, the *additional information index* (*ai*) for D is defined as follows:

$$ai(D) = \frac{\sum_{i=1}^{n} additionalinf(d_i)}{n}$$

Evaluation of Similar Knowledge

In domains such as biomedicine, it is common to find ontologies that, besides containing a particular concept, provide many other similar concepts. When assessing the semantic richness that an ontology provides for a context, it is also important to consider these similar concepts.

Example 6. Suppose that it is necessary to find the best ontology to represent the term *heart*. If we consider the example ontologies E and F from Figure 4.3 we can see that both cover the term, because both contain the concept *heart*. However, it is clear that the ontology F provides some additional knowledge related to the concept *heart* that can be useful. This ontology contains other 6 concepts whose meaning belongs to the domain of cardiology, while the ontology E contains none.

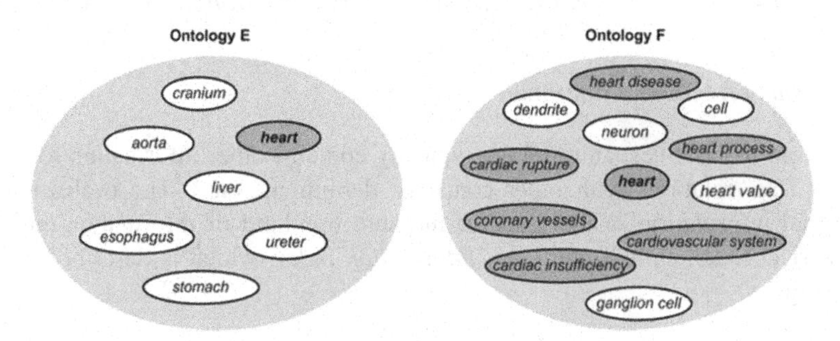

Fig. 4.3 Sets of concepts for two example ontologies

The evaluation of similar knowledge is addressed to measure how much knowledge contains an ontology, related to the meaning of a specific concept. It is done by calculating a *similar knowledge index*, which is explained as follows.

Definition 5 (*similar knowledge index*). Suppose that C is the set of concepts in an ontology. The function similar knowledge for a concept $c \in C$ returns the number of concepts in the ontology whose name contains the name of c or any of its synonyms. Considering that $substrings(s)$ is the function that returns all the possible substrings of a given string s, then:

$$similarknowledge(c) = \\ |\{c' \in C | c' \neq c \wedge (expand(conceptname(c)) \cap substrings(conceptname(c')) \neq \varnothing)\}|$$

Example 7. Considering the ontologies E and F from Figure 4.3, suppose that c_1 is the concept *heart* in the ontology E, and c_2 is the concept *heart* in the ontology F. Also suppose that $expand(heart) = \{heart, cardio, cardiac, coronary\}$. Then:

$$similarknowledge(c_1) = |\{\}| = 0$$
$$similarknowledge(c_2) = |\{heart\ disease, cardiac\ rupture, heart\ process, \\ coronary\ vessels, cardiovascular\ system, cardiac\ insufficiency\}| = 6$$

Suppose that o is an ontology whose set of concepts is C. Also suppose that $D = \{d_1, d_2, \ldots, d_n\} \subseteq C$ is the subset of concepts in o that cover the input terms. Then, the *similar knowledge index (si)* for D is defined as:

$$si(D) = \frac{\sum_{i=1}^{n} similarknowledge(d_i)}{n}$$

Semantic Richness Score

After carrying out the evaluation of relatives, additional information and similar knowledge that an ontology provides for a given context, these three measures are combined into a single value that represents the semantic richness of the ontology in such context. We propose a metric, called *Semantic Richness Score (SRscore)*, which is defined as follows:

Definition 6 (*SRscore*). Let $T = \{t_1, t_2, \ldots, t_n\}$ be the set of initial terms after the steps of normalization and correction, with $n = |T|$. Suppose that o is an ontology whose set of concepts is C. Also, suppose that $D = \{d_1, d_2, \ldots, d_m\} \subseteq C$ is a subset of concepts from o, each of which covers one term in T. Then the function *SRscore* is expressed as:

$$SRscore(D) = w_r \cdot norm_{ri}(ri(D)) + w_a \cdot norm_{ai}(ai(D)) + w_s \cdot norm_{si}(si(D))$$

Where w_r, w_a and w_s are weight factors, such that $w_r + w_a + w_s = 1$. These weights allow to give more or less importance to each semantic richness criterion. The function *norm* is a normalization function, which will be explained in Section 4.3.3.

4.3.2.3 Popularity Evaluation

Apart from assessing how well each ontology covers a specific context and the richness of its structure, there is another aspect that requires special attention: What about the relevance of the ontology in the community? According to the definition of ontology provided by Studer [15], *"an ontology captures consensual knowledge, that is, it is not private to some individual, but accepted by a group"*. With this in mind, any method addressed to evaluate an ontology should also take into account the level of collective acceptability or popularity of the ontology.

The presented approach proposes to evaluate the popularity of an ontology by using the collective knowledge stored in widely-known Web 2.0 resources, whose value is created by the aggregation of many individual user contributions. It consists on counting the references to the ontology from each resource. Depending of the kind of resource, the criteria to measure the references will be different. For example, in a resource like Wikipedia, we could count the number of results obtained when searching for the name of the ontology; in Del.icio.us, we could measure the number of users who have added the URL of the ontology to their favorites; in Google Scholar or PubMed, retrieving the number of papers which reference the ontology; or even in Twitter, by checking the number of twitts dedicated to the ontology. Apart from these general-purpose resources, it could also be taken into account Web 2.0 resources aimed to cover specific fields, like the NCBO Bioportal[3] for the biomedical domain. According to this idea, we propose to assess the popularity of an ontology on the basis of the following metric.

Definition 8 (*Pscore*). Let O be the set of all ontologies, $R = \{r_1, \ldots, r_n\}$ is the set of Web 2.0 resources taken into account and $refs : O \times R \rightarrow N$ is the function that returns the number of references to an ontology $o \in O$ from a resource $r \in R$ according to a predefined criterion. Then:

$$Pscore(o,R) = \sum_{i=1}^{n} w_i \cdot norm_{ri}(refs(o,ri))$$

where w_i is a weight factor and $norm$ is a normalization function, which will be explained in Section 4.3.3.

Example 8. Suppose that we are evaluating the popularity of Gene Ontology[4], which is a well-known ontology about gene products, and that we just want to take into account the resources Wikipedia and PubMed. At the moment of writing this paper, in the English Wikipedia[5] there are 84 content pages with the text "Gene Ontology", and in PubMed there are 3039 biomedical articles that contain such text. Supposing that $norm_{wikipedia}(84) = 1$ and that $norm_{pubmed}(3039) = 0.75$ then, the *Pscore*

[3] http://bioportal.bioontology.org
[4] http://www.geneontology.org
[5] http://en.wikipedia.org

would be calculated according to the following expression (with $w_{wikipedia} = 0.5$, $w_{pubmed} = 0.5$):

$$PScore = 0.5 \cdot norm_{wikipedia}(84) + 0.5 \cdot norm_{pubmed}(3039) = 0.5 + 0.5 \cdot 0.75 = 0.875$$

4.3.3 Normalization

The proposed approach requires a normalization function, which transforms a value in the interval $[0, +\infty)$ to the interval $[0, 1]$. This normalization can be achieved using different techniques. One of the most simple options could be to achieve a linear normalization (e.g. dividing by the maximum value), but this could cause the loss of differentiating information if the sample is not uniformly distributed. We propose to use a normalization function able to retain the discrimination capacity of the original values, which is defined as follows.

Definition 7 (*normalization*). Suppose that S is a sample of values of a specific feature (e.g. number of references to the ontology from Wikipedia), in the interval $[0, +\infty)$, for a reference set of ontologies O. An equal-frequency discretization strategy [16] is applied to the values in S, so that the continuous range of the feature is divided into k intervals that have equal frequency. The number of intervals is calculated according to the Sturges rule [17]. Let $I = \{I_1, I_2, \ldots, I_k\}$ be this set of intervals. Then, the normalization function $norm : [0, +\infty) \rightarrow [0, 1]$ is defined as:

$$norm(x) = \begin{cases} 0, & \text{if } x = 0 \\ i/k & \text{if } x \in I_i \end{cases}$$

Example 9. Suppose that $O = \{o_1, o_2, \ldots, o_{10}\}$ is a set of 10 ontologies, and that $S = \{4, 21, 2, 0, 3215, 0, 3, 1, 6, 47\}$ is the number of references from Wikipedia to each respective ontology in O (e.g. the ontology o_2 has 21 references from Wikipedia). According to the Sturges rule ($1 + 3.22 \cdot \log(n)$, with $n = 10$), this sample would be divided into 4 intervals of equal frequency. In this case, the separation points between intervals would be calculated as the percentiles 25, 50 and 75 (i.e. quartiles). This points are $\{2.5, 5, 34\}$. The normalization function which would allow to transform any number of references from Wikipedia to the range $[0, 1]$ would be:

$$norm(x) = \begin{cases} 0, & \text{if } x = 0 \\ 1/4 = 0.25, & \text{if } 0 < x < 2.5 \\ 2/4 = 0.5, & \text{if } 2.5 \le x < 5 \\ 3/4 = 0.75, & \text{if } 5 \le x < 34 \\ 4/4 = 1, & \text{if } 34 \le x \end{cases}$$

4.3.4 Final Ontology Score

The final score for an ontology is a value in the interval $[0,1]$, which is calculated once the three previously explained measures have been applied for each candidate ontology. It is calculated by aggregating the obtained values, according the following definition:

Definition 8 (***final score***)**.** Let $T = \{t_1, t_2, \ldots, t_n\}$ be the set of initial terms after the steps of normalization and correction. Suppose that o is an ontology whose set of concepts is C. Also, suppose that $D = \{d_1, d_2, \ldots, d_m\} \subseteq C$ is a subset of the concepts in o, each of which covers one term in T. Then:

$$finalscore(o,T) = w_{CC} \cdot CCscore(o,T) + w_{SR} \cdot SRscore(D) + w_P \cdot Pscore(o)$$

where w_{CC}, w_{SR} and w_P are the weights assigned to the *CCscore*, *SRscore* and *Pscore*, respectively, such that $w_{CC} + w_{SR} + w_P = 1$. Finally, the candidate ontologies are ranked according to their final score, and such ranking constitutes the result of the ontology recommendation process.

4.4 Evaluation and Results

On the basis of the proposed approach, we have built a prototype of an ontology recommender system for the biomedical domain and carried out an experiment to test the validity of the approach. We decided to choose the specific field of anatomy for the evaluation. As explained in [6], focusing on ontologies covering very similar domains is more challenging than searching through ontologies that represent different domains. It will be much easier to filter out an ontology about, for example, transport when searching for terms belonging to the domain of anatomy than to filter between overlapping biomedical ontologies.

The selected spell checker to correct possible typographical mistakes was the Yahoo Spelling Suggestion service[6]. As the terminological resource to achieve the semantic expansion, we have chosen the Unified Medical Language System (UMLS). We used a set of initial terms composed by 80 terms randomly chosen from the area of anatomy. We also built a repository of candidate ontologies, composed by 200 biomedical ontologies, containing a total of more than 1.860.000 concepts. In order to test the system's ability to discriminate ontologies that are not popular in the biomedical community, we also added to the repository an ontology constructed by hand, which we named "Example Anatomy Ontology". This ontology is composed by 81 concepts: 80 concepts covering the 80 initial terms (and therefore, it provides the maximum context coverage), and the concept *Thing* as the parent of the other concepts. As the Web 2.0 resources, we have selected Wikipedia (counting the number of content pages containing the name of the ontology), PubMed (articles

[6] http://developer.yahoo.com/search/web/V1/spellingSuggestion.html

containing the ontology name), Twitter (*twitts* containing the ontology name) and BioPortal (ontology name listed in BioPortal or not).

The evaluation of the approach also required to determine the set of weights that allowed to give more or less importance to each parameter. This adjustment was based on the opinion of 5 experts in biomedical ontologies, which have expressed their view about the values to be used by filling a short questionnaire. The English version of this questionnaire is available at `http://tinyurl.com/6c88bnc`. The complete list of initial terms, the list of ontologies in the repository and the set of weights used, are available at `http://tinyurl.com/48onpdf`. The prototype provided the results that are shown in Table 4.2 and Figure 4.4 (only top 10 recommended ontologies are shown).

Table 4.2 Results of the ontology recommendation process

Pos.	Ontology	*CCscore*	*SRscore*	*Pscore*	Final score
1	Foundational Model of Anatomy	0.825	0.850	0.453	**0.761**
2	NCI Thesaurus	0.625	0.934	0.863	**0.751**
3	Medical Subject Headings	0.513	0.817	1.000	**0.684**
4	Cardiac Electrophysiology Ontology	0.825	0.562	0.280	**0.653**
5	University of Washington Digital Anatomist	0.700	0.883	0.100	**0.634**
6	Example Anatomy Ontology	1.000	0.024	0.000	**0.556**
7	Logical Observation Identifier Names and Codes	0.263	0.967	0.697	**0.528**
8	Galen Ontology	0.538	0.271	0.280	**0.419**
9	BRENDA tissue / enzyme source	0.313	0.746	0.280	**0.419**
10	CRISP Thesaurus	0.363	0.683	0.180	**0.411**

On examining the results, it is possible to highlight the following observations: (i) the top 10 ontologies (except the example ontology) are widely known biomedical ontologies; (ii) the Foundational Model of Anatomy is the most recommended ontology, providing a 82% of coverage for the initial terms. The high context coverage provided by this ontology causes it to be above other ontologies that are richer and more popular, like the NCI Thesaurus or Medical Subject Headings; (iii) the top 10 contains other two ontologies directly related to the domain of anatomy, which also provide a good context coverage: the Cardiac Electrophysiology Ontology (82%) and the University of Washington Digital Anatomist (70%); (iv) in spite of that the example ontology (Example Anatomy Ontology) provides a maximum context coverage (100%), it is ranked at position 6, due to its low levels of semantic richness and popularity.

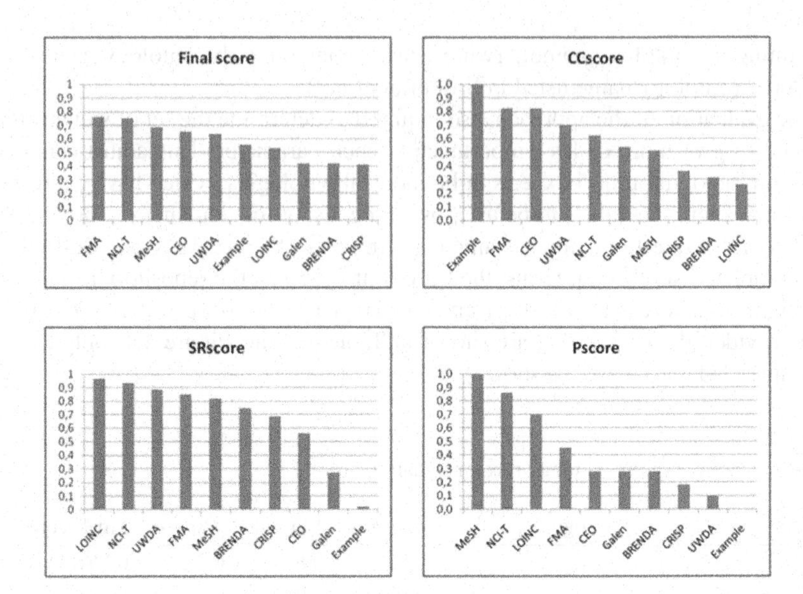

Fig. 4.4 Rankings obtained according to the final score, *CCscore*, *SRscore* and *Pscore*

4.5 Conclusions and Future Work

Automatic ontology recommendation is a crucial task to enable ontology reusing in emerging knowledge-based applications and domains like the upcoming Semantic Web. In this work, we have presented an approach for the automatic recommendation of ontologies that is based on evaluating each ontology according to three different aspects: (i) how well the ontology represents a given context, (ii) the semantic richness of the ontology in such context and (iii) the popularity of the ontology. Results show that the adequate combination of these criteria allows to obtain a valid recommendation, and that the Web 2.0 is a good resource to obtain the collaborative knowledge required to the evaluation of ontology popularity.

As an immediate future work, we plan to develop strategies to automatically identify the best weights for a given problem, instead of having to adjust their values by hand. Investigating new methods to evaluate ontology popularity is another future direction. We also have interest in studying if reasoning techniques can help to improve our scoring strategies.

Acknowledgements. Work supported by the Carlos III Health Institute (grant FIS-PI10/02180), the "Galician Network for Colorectal Cancer Research" (REGICC, Ref. 2009/58), funded by the Xunta de Galicia, and the "Ibero-NBIC Network" (209RT0366) funded by CYTED.

References

1. Sabou, M., Lopez, V., Motta, E., Uren, V.: Ontology Selection: Ontology Evaluation on the Real Semantic Web. In: Evaluation of Ontologies on the Web Workshop, Held in Conjunction with WWW 2006, Edinburgh, Scotland (2006)
2. Gómez-Pérez, A.: Some Ideas and Examples to Evaluate Ontologies. In: 11th IEEE Conference on Artificial Intelligence Applications, pp. 299–305. IEEE Computer Society Press, Los Angeles (1995)
3. Gómez-Pérez, A.: From Knowledge Based Systems to Knowledge Sharing Technology. In: Evaluation and Assessment. KSL Lab, Stanford University, CA (1994)
4. Berners-Lee, T., Hendler, J., Lassila, O.: The semantic Web. Scientific American 284(5), 34–43 (2001)
5. Supekar, K., Patel, C., Lee, Y.: Characterizing Quality of Knowledge on Semantic Web. In: Seventeenth International FLAIRS Conference, Miami, Florida, USA, pp. 220–228 (2004)
6. Alani, H., Noy, N., Shah, N., Shadbolt, N., Musen, M.: Searching Ontologies Based on Content: Experiments in the Biomedical Domain. In: Fourth International Conference on Knowledge Capture (K-Cap), Whistler, BC, Canada, pp. 55–62. ACM Press (2007)
7. Alani, H., Brewster, C., Shadbolt, N.R.: Ranking Ontologies with aKTiveRank. In: Cruz, I., Decker, S., Allemang, D., Preist, C., Schwabe, D., Mika, P., Uschold, M., Aroyo, L.M. (eds.) ISWC 2006. LNCS, vol. 4273, pp. 1–15. Springer, Heidelberg (2006)
8. Netzer, Y., Gabay, D., Adler, M., Goldberg, Y., Elhadad, M.: Ontology Evaluation through Text Classification. In: Chen, L., Liu, C., Zhang, X., Wang, S., Strasunskas, D., Tomassen, S.L., Rao, J., Li, W.-S., Candan, K.S., Chiu, D.K.W., Zhuang, Y., Ellis, C.A., Kim, K.-H. (eds.) WCMT 2009. LNCS, vol. 5731, pp. 210–221. Springer, Heidelberg (2009)
9. Vilches-Blázquez, L., Ramos, J., López-Pellicer, F., Corcho, O., Nogueras-Iso, J.: An Approach to Comparing Different Ontologies in the Context of Hydrographical Information. In: Heidelberg, S.B. (ed.) Information Fusion and Geographic Information Systems. Lecture Notes in Geoinformation and Cartography, vol. 4, pp. 193–207. Springer, Berlin (2009)
10. Jonquet, C., Musen, M., Shah, N.: Building a Biomedical Ontology Recommender Web Service. Journal of Biomedical Semantics (S1), 1–18 (2010)
11. Sabou, M., Lopez, V., Motta, E.: Ontology Selection for the Real Semantic Web: How to Cover the Queen's Birthday Dinner? In: Staab, S., Svátek, V. (eds.) EKAW 2006. LNCS (LNAI), vol. 4248, pp. 96–111. Springer, Heidelberg (2006)
12. Brank, J., Grobelnik, M., Mladenic, D.A.: Survey of Ontology Evaluation Techniques. In: Conference on Data Mining and Data Warehouses (SiKDD 2005), Ljubljana, Slovenia (2005)
13. Jones, M., Alani, H.: Content-based ontology ranking. In: 9th Int. Protégé Conference, Stanford, CA (2006)
14. Romero, M.M., Vázquez -Naya, J.M., Munteanu, C.R., Pereira, J., Pazos, A.: An Approach for the Automatic Recommendation of Ontologies Using Collaborative Knowledge. In: Setchi, R., Jordanov, I., Howlett, R.J., Jain, L.C. (eds.) KES 2010. LNCS, vol. 6277, pp. 74–81. Springer, Heidelberg (2010)
15. Studer, R., Benjamins, V.R., Fensel, D.: Knowledge Engineering: Principles and Methods. IEEE Transactions on Data & Knowledge Engineering 25(1-2), 161–197 (1998)
16. Liu, H., Hussain, F., Tan, C., Dash, M.: Discretization: An Enabling Technique. Data Mining and Knowledge Discovery 6(4), 393–423 (2002)
17. Daniel, W., Wayne, W.: Biostatistics: a Foundation for Analysis in the Health Sciences, 9th edn. John Wiley and Sons, New York (2009)

Part III
Trust & Recommendation

Chapter 5
Implicit Trust Networks: A Semantic Approach to Improve Collaborative Recommendations

Manuela I. Martín-Vicente, Alberto Gil-Solla, and Manuel Ramos-Cabrer

Abstract. Collaborative recommender systems suggest items each user may like or find useful basing on the preferences of other like-minded individuals. Thus, the main concern in a collaborative recommendation is to identify the most suitable set of users to drive the selection of the items to be offered in each case. To distinguish relevant and reliable users from unreliable ones, trust and reputation models are being increasingly incorporated in these systems, by using network structures in which nodes represent users and edges represent trust statements. However, current approaches require the users to provide explicit data (about which other users they trust or not) to form such networks. In this chapter, we apply a semantic approach to automatically build implicit trust networks and, thereby, improve the recommendation results transparently to the users.

5.1 Introduction

Recommender systems [10, 1] arose to save the users from dealing with the overwhelming amounts of information present in numerous domains (such as TV, tourism, e-commerce, etc.), by providing personalized suggestions. That is, these systems aim to select and offer to each user, from among the myriad of items available, those that he/she may like or find useful. For that purpose, from their inception, recommender systems rely on *user profiles* that typically store the past ratings records of the users. Indeed, the profile of the user who will receive the current recommendation (hereafter, *active user*), together with the *items database* to recommend (e.g. audiovisual contents in TV, tourist packages in tourism, commercial products in e-commerce, etc.), make up the basic information upon which every recommendation process is grounded –see Figure 5.1. Besides this common

Manuela I. Martín-Vicente · Alberto Gil-Solla · Manuel Ramos-Cabrer
SSI Group, Department of Telematic Engineering, University of Vigo, 36301 Vigo, Spain
e-mail: {mvicente,agil,mramos}@det.uvigo.es

J.J. Pazos Arias et al.: Recommender Systems for the Social Web, ISRL 32, pp. 107–119.
springerlink.com

information, different additional data is required depending on the personalization strategy utilized by the system.

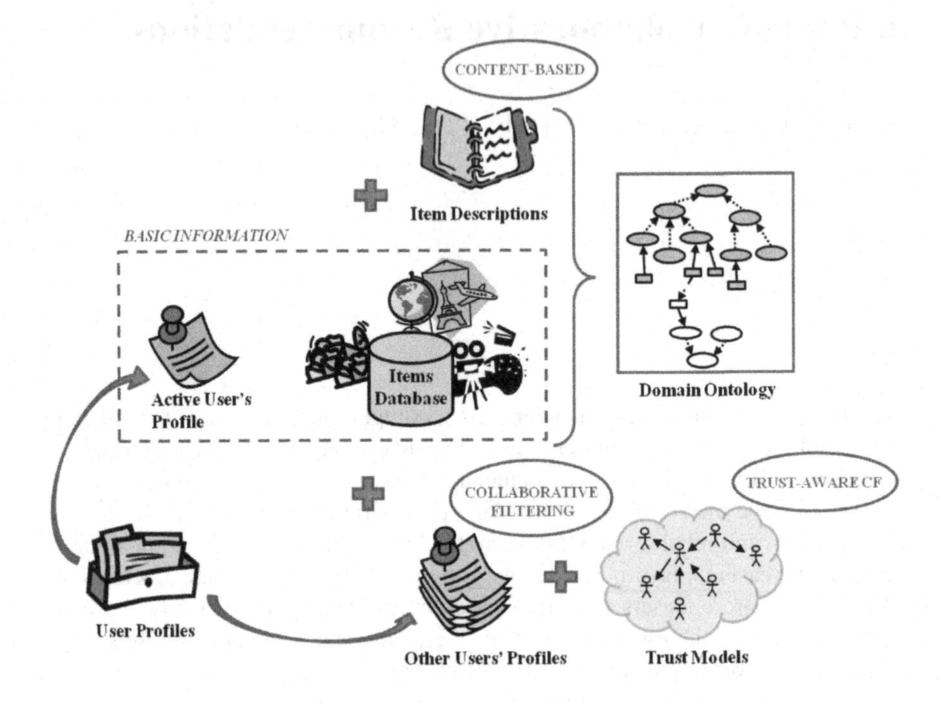

Fig. 5.1 Recommender systems: strategies and their required data.

The first recommendation strategy was *content-based filtering* [18, 13], characterized by suggesting to the active user items similar to those he/she liked in the past. As shown in Figure 5.1, the additional data used by content-based recommenders are the *item descriptions*, which are explored to determine the similarity between the items to recommend and the ones stored in the profile of the active user. The main problem with this technique is its tendency to overspecialization (only suggesting items very similar to those the user already knows), which caused it to fall into disuse in favor of *collaborative filtering* strategies [3, 15, 12, 8] –since then, it has only been adopted in *hybrid approaches* [2, 5, 14, 4, 17] to complement the latter. Collaborative filtering fights overspecialization offering to the active user items that were appealing to other individuals with similar preferences (the so-called *neighbors*). The additional data these strategies need are the *other users' profiles*, which are compared to the active user's in order to form his/her neighborhood –traditionally, selecting those who have assigned the most similar ratings to the same items. Then, to predict the interest in an item the active user does not know, collaborative filtering techniques consider the ratings his/her neighbors have given to that item.

However, traditional collaborative recommenders neglect the fact that, relying solely on the ratings registered in the users' personal profiles, recommendation results can be distorted by malicious or unreliable users. For that reason, approaches incorporating *trust models* into collaborative recommender systems have recently gained momentum. These proposals complement (or even replace) the similarity-based neighborhood formation with a trust-aware selection of neighbors. The idea is to carry out the neighborhood formation considering not only if the candidates share similar rating records with the user who will receive the recommendation, but also whether they are trustworthy or not. In these approaches, the reliability of a candidate is usually obtained by using a trust network, in which nodes represent users and edges stand for trust statements. Nevertheless, most current methods rely on explicit trust networks, in which users are required to state the trust data (indicating how much they trust other users), and therefore they cannot be applied to open systems where individuals may not know each other.

In this chapter, we make use of semantics to build *implicit trust networks* in order to incorporate trust and reputation in the neighborhood formation with no request of explicit trust data, thereby improving collaborative recommender systems transparently to the users. With this goal in mind, we include a new piece in the puzzle of Figure 5.1: an *ontology*. As depicted in the figure, this element is nothing more than a formal representation of the *items database* and the *item descriptions*. In general terms, an ontology describes hierarchically organized concepts of a particular domain and the relationships between them (which are represented by classes and properties, respectively) and it is populated by including specific instances of both entities. For example, if considering the e-commerce domain, the ontology will formalize the semantic annotations of the available commercial products for recommendation. In that case, the book entitled *Murder on the Orient Express* could be categorized as an instance of the class *Mystery Novels* (in the *Books* hierarchy), with attributes like *Agatha Christie* (instance of the attribute class *Writers*). By using an ontology, reasoning techniques can be applied to discover new knowledge and a great amount of useful data can be included in the personalization process. Our strategy infers trust relations between pairs of users from the results in previous recommendations. Exploiting the semantics, we can estimate how reliable a user is when contributing (as part of a neighborhood) to recommendations of different kind of items, at any level of the ontology hierarchy: from the leaf level, in which classes represent the most specific concepts, to the category level, in which classes (subclasses of the root of the hierarchy) represent the most generic ones.

The chapter is organized as follows: Firstly, Section 5.2 discusses related work in this area. Next, Section 5.3 focuses on the trust model included in our system, by describing the semantics-based strategy to build our implicit trust network and how the reputation values that we get from it influence the neighborhood formation. Lastly, Section 5.4 concludes the chapter and points out the improvements achieved with our approach.

5.2 Related Work

Trust and reputation are factors that influence decision-making across the board, so as to distinguish relevant and reliable sources of information from unreliable ones. Recent research publications deal with the issue of how to use trust and reputation in recommender systems to improve their accuracy and reliability.

The increasing interest in issues of trust has raised disagreement with not only its application and relevance, but even its basic definition. While Kwon et al. in [7] differentiate trustworthiness from similarity as separate, independent key dimensions of the credibility of an information source, Ziegler and Golbeck conclude in [20] that there is correlation between interpersonal trust and interest similarity. On the other hand, while some approaches do not make a distinction between trust and reputation, and only use one of the two concepts in their models, the work in [6] distinguishes trust from reputation systems, defining trust as an individual measure based only on first-hand experience, and reputation as a collective measure of trustworthiness based on the referrals or ratings from members in a community.

Regarding how to incorporate trust models during the recommendation process, a large number of researchers make use of trust data provided explicitly by the users. Particularly, it is usual to employ data from Epinions. Epinions is an e-commerce site where users can rate products and write reviews about them. Also, users can assign trust ratings to reviewers based on the degree to which they have found them helpful and reliable in the past. In other words, users are allowed to express their degree of trust in other users. An example of the use of this web site can be found in [9]. In that work, Massa and Avesani build a trust model directly from trust data provided by users as part of Epinions, and use it to compute a relevance measure to be used as a complementary weight (in addition to user similarity) in the recommendation process. In [19], Zhang et al. aim to get a more accurate trust management and infer, from the general trust relationships available in Epinions, trust relationships in certain categories. For its part, the work presented in [16] relies on data from the aforesaid e-commerce site to deal with a recurring problem in recommender systems: the lack of information about the users at the very start. As mentioned in Section 5.1, the main weakness of all these approaches is the requirement of an explicit trust network, which limits their use to individuals that know each other to a greater or lesser extent and are willing to provide trust statements. Our strategy incorporates a trust model into the recommendation process transparently to users, and therefore it can be applied to open systems where individuals may not know each other.

As for the procurement of implicit trust data from the results in previous recommendations, the work most closely related to ours is that of O'Donovan and Smyth [11]. In a like manner, their strategy automatically infers trust and reputation from users' contributions to recommendations. However, they define only two distinct levels of trust: item-level, with respect to the recommendation of specific items (e.g. John's trustworthiness for recommending a Toyota Corolla), and profile-level, representing trust as a whole (e.g. John's trustworthiness). Item-related trust values lead to a *sparsity problem*, since the huge amount of commercial products available makes it very unlikely to benefit from such specific data. Consider, for instance, the search for users with a good record of contributions to recommendations of a certain brand of green tea. At the opposite extreme, user-related trust values lead to an *overgeneralization*, since users can be trustworthy for recommending some types of items but not others. Our work employs semantic reasoning to deal with the aforementioned problems. Semantics allows us to consider different degrees of abstraction, thus providing great flexibility to the process. Continuing the last example, we can take into account, instead of a certain brand of green tea, contributions when recommending any brand of green tea, any kind of tea, any kind of infusions, any kind of beverages, and so on. The greatest level of abstraction our strategy would consider in this case is foodstuff, in order to avoid overgeneralization.

5.3 Implicit Trust-Aware Neighborhood Selection

This section describes how we incorporate trust and reputation models in the neighborhood formation transparently to the users, with no request of explicit trust data. As depicted in Figure 5.2, we can distinguish four steps in the process. In step 1 (presented in Section 5.3.1) we exploit the semantics to infer higher quality trust relations between pairs of users, reasoning from a record of past recommendations results. The step 2 (detailed in Section 5.3.2) involves gathering together those relationships to form an implicit trust network among all the users of the system. Then, in step 3 (put forward in Section 5.3.3) our strategy explores the resulting network to obtain users' reputation measures. Finally, during the step 4 (illustrated at the bottom of the figure, and described in Section 5.3.4) we incorporate trust and reputation values in the neighborhood selection process.

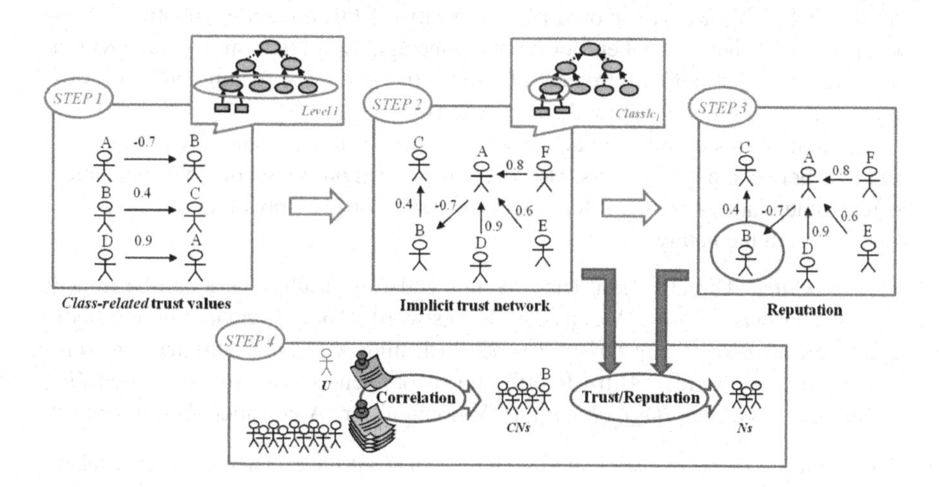

Fig. 5.2 Overview of our strategy.

5.3.1 A Semantic Method for Mining Trust Relations

Our strategy starts from the results in previous recommendations to infer trust relations between pairs of users. As mentioned in Section 5.1, in collaborative recommenders each recommendation process involves a certain group of users of the system, one of them playing the role of active user and the others playing the role of neighbors. Since that group varies from process to process, with the passage of time a given user can have performed one or both roles in the recommendations in which he/she has participated. In order to infer the trustworthiness of a user, we need to focus on his/her contributions as a neighbor. In particular, we quantify positively or negatively a trust relation between two users depending on whether the one has contributed (as a neighbor) to recommendations that have been appealing or unappealing to the other (as active user). Graphically, we represent a trust relation between two users A and B by a link from A to B, stating how much A trusts or distrusts B –as a result of B's contributions in recommendations to A (see step 1 in Figure 5.2).

But, what determines a neighbor's contribution to be a success or a failure? Specifically, we consider the contribution of a neighbor N in the recommendation of the item i to an active user U a success if the difference between the degree of interest (DOI[1]) showed by U after that recommendation and the one of N is less than a given error bound, according to the expression in Eq. 5.1.

$$E_N i, U = |DOI_N(i) - DOI_U(i)| = \varepsilon \in [0,2] \quad (5.1)$$

Note that we isolate N's contribution from the overall recommendation, quantifying whether or not it had a positive or negative effect on the final recommendation independently of the other neighbors. This way, we are able to detect N's successes not only when he/she has favored (with a positive DOI) a recommendation that is accepted by U, but also when he/she has gone against a recommendation (with a negative DOI) that is finally rejected by the active user –the recommendation was made due to the other neighbors' contributions.

These measures of success/failure serve as a basis to automatically obtain trust relations between pairs of users. However, as pointed out in Section 5.2, inasmuch as recommendations usually refer to specific items, we can only obtain two types of trust values immediately:

- *Item-related*: Directly from each recommendation results, there can be mined trust relations such as "A considers B trustworthy for recommending the book entitled *Murder on the Orient Express*" (dealing with an e-commerce recommender), "A considers B trustworthy for recommending the movie entitled *The Good, the Bad and the Ugly*" (in a TV domain), or "A considers B trustworthy

[1] We define the *degree of interest* of a user in an item in order to exploit not only the explicit ratings provided by the users (which can be scarce) but also the implicit feedback that can be obtained from their interaction with the system –e.g monitoring whether or not the active user purchases the product (in e-commerce), or whether or not he/she watches the recommended program (in the TV domain).

for recommending the activity *Diving in the Great Barrier Reef*" (from a tourism recommender). The downside is that item-related trust values lead to a *sparsity* problem, because trying to select trustworthy neighbors for recommending one specific item (from among the myriad of items available) can be like looking for a needle in a haystack.

- *User-related*: Generalizing from all the contributions to recommendations we can get trust relations like "A trusts B", but these relations imply an *overgeneralization*, as it is very unlikely for a user to be trustworthy for recommending all different kinds of items available for recommendation in a given domain. For instance, considering the domains we have used as examples, a user can be trustworthy for recommending: (i) *Books* but not *Music* or *Sports Equipment* (in e-commerce), (ii) *Fiction* programs but not *Sports* (in TV), and (iii) *Adventure* and *Cultural* tourist packages but not *Religious* ones (in tourism).

To deal with these problems, we rely on a semantic approach. By exploiting the semantics formalized in an ontology, a great amount of useful knowledge can be incorporated into the process, which leads to higher quality trust relations. Indeed, whereas a syntactic approach is limited to obtain relations only attending to the words that identify a certain item (in the ongoing examples: "*Murder on the Orient Express*", "*The Good, the Bad and the Ugly*" and "*Diving in the Great Barrier Reef*"), semantics enables reasoning about the essence of those words (e.g. recognizing: *Murder on the Orient Express* as a mystery novel book, a commercial product; *The Good, the Bad and the Ugly* as a western movie, a TV content; and *Diving in the Great Barrier Reef* a water activity, an adventure tourism package). In our strategy, we explore the semantics of users' preferences to infer *class-related* trust values at any level of the ontology hierarchy.

- Firstly, we obtain trust values at the leaf level, from past recommendations of products classified under any leaf class. Specifically, the value of a trust relation from an active user U to a neighbor N with regard to a leaf class C_L comes from the average of N's successful and unsuccessful contributions to U in recommendations of products classified in the ontology under that class, as shown in Eq. 5.2. In this expression, E_Ni,U is the measure of N's success/failure in the recommendation of product i to U, and M is the number of recommendations to user U of products belonging to class C_L in which the neighbor N has participated.

$$T_{U,N}(C_L) = \frac{1}{M}\sum_{i=1}^{M}(1 - E_Ni, U) \tag{5.2}$$

By way of illustration, consider the simplification of an e-commerce ontology hierarchy depicted at the left-hand side of Figure 5.3. In such a scenario, we can infer whether or not "A considers B trustworthy for recommending mystery novels" from B's successes/failures in recommendations to A of products belonging to the class *Mystery* in the ontology (such as *Murder on the Orient Express*, *The Maltese Falcon*, etc.).

- Afterwards, we get trust values at any lower level by propagating leaf-class values up through the hierarchy. In the current example, the next level corresponds to the class *Novels* and it will gather B's contributions when recommending novels of any genre. Our proposal adopts the approach proposed by Ziegler [21] (applied here to trust values rather than levels of interest) to propagate the values, therefore taking into account both the level of the superclass being considered and the number of siblings that the class which spreads its value has. As it is shown in Eq. 5.3, the highest values will be those of the more specific classes in the hierarchy, i.e., the closest to the leaf classes. Likewise, the trust value of a superclass C_{p+1} will be more significant the greater the value spread by the class C_p (subclass of C_{p+1}) and the lesser the number of siblings such class has.

$$T_{U,N}(C_{p+1}) = \frac{T_{U,N}(C_p)}{1 + \#sib(C_p)} \tag{5.3}$$

  As every superclass can have more than one subclass, and therefore get more than one spread value, we calculate its total trust value by averaging out the values spread by each one of the classes that belong directly to it –obtained through Eq. 5.3.
- In the extreme case, with level-one classes, we infer trust values related to a category of items. In the present example, the category which we reach is represented by the class *Books*, pictured at the left-hand side of Figure 5.3. Thus, we obtain relations like "A considers B trustworthy for recommending books". Likewise, we will get trust values related to the rest of the categories in the ontology (e.g. *Music*, *Sports Equipment*, *Foodstuff*, etc., in the e-commerce domain).

5.3.2 Building an Implicit Trust Network

From the trust relations inferred between pairs of users, we build a trust network among the users of our system. The idea is to organize the one-to-one relations into a structure that interconnects all the users, who become the nodes of the resulting network (see Figure 5.2). Once the network is formed, to identify the roles a user have performed in past recommendation processes (active user, neighbor, or both) all we have to do is to look at the directions of the links connected to his/her node: the links pointing at it represent assessments of the user's role as neighbor, whereas the links starting from it are due to his/her role as active user. For example, in step 2 of Figure 5.2 we can see that B has participated as neighbor in recommendations to A, and also he/she has been the active user of recommendations in which C has taken part.

Notwithstanding, the building process is not that straightforward and, as we will discuss below, implies a trade-off. The reason is that the resulting network will be diverse depending on which level of class-related trust relations we choose to make it up. That is to say, we are able to form as many types of implicit trust networks as levels of abstraction are in the ontology hierarchy. The differences stem from the links existing among the nodes and the number of layers that make up the network.

On the one hand, relations inferred from classes at one or another level bring about different connections between the nodes. On the other hand, there will be as many layers in the network as classes are defined in the ontology at the level that is being considered.

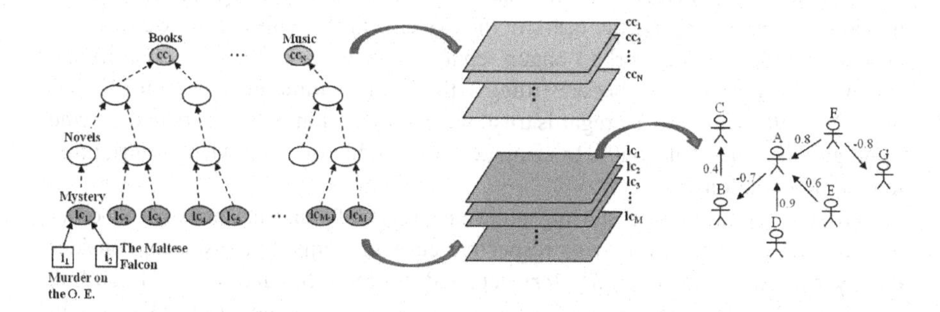

Fig. 5.3 Implicit trust network formation.

- If considering the leaf level, the network will have as many layers as leaf classes are defined in the ontology. Each layer represents the relationships set up among the users according to a certain leaf class. Figure 5.3 depicts how each leaf class (lc_1 to lc_M) of the ontology (in this case, an ontology of commercial products) corresponds to a layer in the network –at the center bottom of the figure. A small sample of one of these layers is shown at the right-hand side of Figure 5.3 containing, for instance, *Mystery*-related trust values on its links.
- At the opposite extreme, if considering the category level, the network will have as many layers as categories (level-one classes) are defined in the ontology (cc_1 to cc_N in Figure 5.3). Each layer is conditioned to a certain category. For example, in the e-commerce domain, a network may have one layer containing *Books*-related trust values on its links, another one containing *Music*-related values, etc.

The number of classes per level in any ontology increases as we descend in the hierarchy. Consequently, the greater the level we choose, the higher the number of layers the network will have ($N \ll M$, in Figure 5.3). This fact has two implications: on the one hand, opting for a high level entails much more computational cost to build and manage the network; on the other hand, a high level will lead to more accurate findings in our trust-aware neighbor selection strategy, since the trust values we will have at our disposal will be related to more specific concepts. Hence, there is a trade-off between computational cost and accuracy. Choosing a certain level to build the network will depend on the available runtime environment: with a limited processing power we will select a lower level; otherwise, the greatest possible.

5.3.3 Our Reputation Measure

Following the definition of *trust* as "how reliable one user considers another", *reputation* can be defined as "how reliable a user is generally considered". So far, we have explained how our strategy infers trust values between pairs of users and how we structure such independent relations into a trust network, organizing the connections set up among all the users in the system. In this subsection, we describe how we explore that network to obtain reputation measures, that is, to quantify the trustworthiness of a given user according to the whole community (see step 3 in Figure 5.2). In the same way as regards trust, we get reputation of the users in their role of neighbors, since our goal is to estimate how reliable they are when contributing to recommendations.

Our strategy takes as a starting point any layer of our trust network in order to define users' reputation on its respective class of items. For instance, consider the layer related to the class *Mystery* depicted in Figure 5.3 and user A. Each link pointing at A states how trustworthy the user that the link comes from (D, E or F in this example) considers A for recommending mystery novels. Thus, A's reputation will state how reliable A is considered in general for recommending mystery novels.

Eq. 5.4 formalizes our definition of reputation for a user (as a neighbor) N, regarding the class C_p. In this equation, L is the number of links pointing at N, and U_i the user from whom the i-th link comes.

$$REPUTATION_N(C_p) = 0.5(1 + \frac{1}{L}\sum_{i=1}^{L} T_{U_i,N}(C_p) \cdot REPUTATION_{U_i}(C_p)) \quad (5.4)$$

Our reputation measure depends on the trust values a user (as a neighbor) gets from others, considering in turn their own reputation. Indeed, we weigh each trust value in relation to the reputation of the user who the link comes from. In this way, values from reliable users are worth more. Then, the greater the positive trust links a user gets and the more trustworthy the users that those links come from are, the greater reputation he/she will have. We can look at the example in Figure 5.3 (right-hand side) to clarify this point. Let us consider user B in this scenario. To obtain B's reputation we need to focus on the links pointing at him/her. In this case, we find only one, from A, with a negative value (A distrusts B). Still, we weigh that value according to A's reputation. A gets three trust values, from D, E and F, all of them positive (they all trust A, to a greater or lesser extent). Consequently, A has high reputation and B gets a very bad reputation measure.

It is worth pointing out that whenever a reputation value is renewed, such change spreads all over the network, updating the corresponding reputation values. Continuing with the example depicted in Figure 5.3, when F's reputation value changes (due to new contributions in collaborative recommendations) it makes A's and G's reputation values to change. Then, the updated reputation value of A leads to modify B's reputation value, and so on.

5.3.4 Incorporating Trust and Reputation in the Neighborhood Formation

The formation of the neighborhood is a key factor when making a recommendation based on collaborative filtering, because it determines the end result of the recommendation to the active user. Below, we put forward how to enhance the neighborhood selection by incorporating the trust values and reputation measures we have presented up to now.

In general, given an active user U, the neighborhood formation strategy consists in the search for the users of the system whose preferences are the most similar to those of U's. Specifically, a vector for each user is created, containing the DOIs of the items recorded in his/her profile; next, the correlation between the vector of the active user and each of the rest of the users is computed; and finally, the highest correlation measures are selected, corresponding to the nearest neighbors to U.

Our strategy, instead of recognizing the resulting group of users as neighbors, considers them as the most suitable candidates to neighbors (*CNs*) for that active user. From among such candidates, we now take into account their trustworthiness. To this aim, we weigh the correlation value obtained between U and each candidate to neighbor with the trust relation between them (if it exists) or the reputation value of the latter –regarding the class or category of the target product for recommendation. That is to say:

- If the candidate to neighbor that is being considered, CN_i, has participated as a neighbor in previous recommendations to U of products belonging to the class or category of products in question (C_p), we can utilize the trust value inferred between them (according to Section 5.3.1), i.e., we can look at the value of the link existing in our implicit trust network from U to CN_i in the corresponding layer. Then, we weigh the correlation value as shown in Eq. 5.5.

$$corr(U,CN_i) \cdot T_{U,CN_i}(C_p) \tag{5.5}$$

- If the candidate to neighbor has never participated in recommendations to U of products classified under C_p in the ontology, there is not a trust value inferred for them. Thus, neither there is a link between them in that layer of our network. In this case, we weigh the correlation value obtained between U and CN_i with CN_i's reputation measure (estimated as presented in Section 5.3.3), as follows:

$$corr(U,CN_i) \cdot REPUTATION_{CN_i}(C_p) \tag{5.6}$$

So, we select as neighbors the candidates who get the highest values of 5.5-5.6. Consequently, a neighborhood will be made up of users who have the most similar preferences to those of the active user and also the most successful history on recommending that kind of items.

5.4 Lessons Learned

In this chapter, we have explored how semantics and trust models can be integrated together into enhanced collaborative recommender systems. The proposed semantics-driven strategy makes the incorporation of trust and reputation in the neighborhood formation process transparently to the users, by obtaining trust relations from a record of results in previous recommendations. Working in tandem with a domain ontology, we are able to reason about such recorded data and discover new knowledge, which allows us to obtain higher quality trust relations. Besides, structuring the relations that are inferred between pairs of users, our strategy builds an implicit trust network among all the users of the system. Through the exploration of the resulting network, we are able to quantify the reputation measure of each user. This procedure enables us to go one step further, by not only taking into account the one-to-one trust relations in the neighborhood selection, but also considering how reliable each candidate to neighbor is according to the whole community.

One of the strengths of our strategy is the generality of its approach: on the one hand, its domain-independent nature makes it possible to be applied in a wide range of recommender systems; on the other hand, the capability of consider different levels of abstraction to build the implicit trust network provides great flexibility to the whole process.

Acknowledgements. This work has been partially funded by the Ministerio de Educación y Ciencia (Gobierno de España) research project TIN2010-20797 (partly financed with FEDER funds), and by the Consellería de Educación e Ordenación Universitaria (Xunta de Galicia) incentives file CN 2011/023 (partly financed with FEDER funds).

References

1. Adomavicius, G., Tuzhilin, A.: Towards the next generation of recommender systems: a survey of the state-of-the-art and possible extensions. IEEE Transactions on Knowledge and Data Engineering 17, 734–749 (2005)
2. Balabanovic, M., Shoham, Y.: Combining content-based and collaborative recommendation. Communications of the ACM 40(3), 1–9 (1997)
3. Burke, R.: Integrating knowledge-based and collaborative filtering in recommender systems. In: Proceedings of the Workshop on AI and Electronic Commerce, pp. 69–72 (1999)
4. Cornelis, C., Lu, J., Guo, X., Zhang, G.: One-and-only item recommendation with fuzzy logic techniques. Information Sciences 177(1), 4906–4921 (2007)
5. Good, N., Schafer, J.B., Konstan, J., Borchers, A., Sarwar, B., Herlocker, J., Riedl, J.: Combining collaborative filtering with personal agents for better recommendations. In: Proceedings of 16th International Conference on Artificial Intelligence, pp. 439–446 (1999)
6. Josang, A., Ismail, R., Boyd, C.: A survey of trust and reputation systems for online service provision. Decision Support Systems 43(2), 618–644 (2007)

7. Kwon, K., Cho, J., Park, Y.: Multidimensional credibility model for neighbor selection in collaborative recommendation. Expert Systems with Applications 36(2), 7114–7122 (2009)
8. Liu, D., Lai, C., Lee, W.: A hybrid of sequential rules and collaborative filtering for product recommendation. Information Sciences 179(20), 3505–3519 (2009)
9. Massa, P., Avesani, P.: Trust-Aware Collaborative Filtering for Recommender Systems. In: Meersman, R. (ed.) OTM 2004. LNCS, vol. 3290, pp. 492–508. Springer, Heidelberg (2004)
10. Montaner, M., Lopez, B., De la Rosa, J.: A taxonomy of recommender agents on the Internet. Artificial Intelligence Review 19(4), 285–330 (2003)
11. O'Donovan, J., Smyth, B.: Mining trust values from recommendation errors. International Journal on Artificial Intelligence Tools 15(6), 945–962 (2006)
12. Oufaida, H., Nouali, O.: Exploiting Semantic Web technologies for recommender systems: a multi view recommendation engine. In: Proceedings of 7th Workshop on Intelligent Techniques for Web Personalization and Recommender Systems, pp. 26–32 (2009)
13. Pazzani, M.J., Billsus, D.: Content-Based Recommendation Systems. In: Brusilovsky, P., Kobsa, A., Nejdl, W. (eds.) Adaptive Web 2007. LNCS, vol. 4321, pp. 325–341. Springer, Heidelberg (2007), `citeseer.ist.psu.edu/resnik99semantic.html`
14. Salter, J., Antonopoulos, N.: CinemaScreen recommender agent: combining collaborative and content-based filtering. IEEE Intelligent Systems 21(1), 35–41 (2006)
15. Su, J., Wang, B., Hsiao, C., Tseng, V.: Personalized rough-set-based recommendation by integrating multiple contents and collaborative information. Information Sciences 180(1), 113–131 (2010)
16. Victor, P., Cornelis, C., De Cock, M., Teredesai, A.M.: Key figure impact in trust-enhanced recommender systems. AI Communications 21(2–3), 127–143 (2008)
17. Wand, R., Kong, F.: Semantic-enhanced personalized recommender system. In: Proceedings of International Conference on Machine Learning and Cybernetics, pp. 4069–4074 (2007)
18. Zenebe, A., Norcio, A.: Representation, similarity measures and aggregation methods using fuzzy sets for content-based recommender systems. Fuzzy Sets and Systems 160(1), 76–94 (2009)
19. Zhang, Y., Chen, H., Jiang, X., Sheng, H., Zhou, L., Yu, T.: Content-based trust mechanism for e-commerce systems. In: Proceedings of the 3rd IEEE Asia-Pacific Services Computing Conference, pp. 1181–1186 (2008)
20. Ziegler, C.-N., Golbeck, J.: Investigating interactions of trust and interest similarity. Decision Support Systems 43(2), 460–475 (2007)
21. Ziegler, C.-N., Schmidt-Thieme, L., Lausen, G.: Exploiting Semantic Product Descriptions for Recommender Systems. In: 2nd ACM SIGIR Semantic Web and Information Retrieval Workshop, pp. 25–29 (2004)

Chapter 6
Social Recommendation Based on a Rich Aggregation Model

Ido Guy

Abstract. This chapter describes a set of recommender systems for both people and content within the enterprise that were all built based on a rich aggregation model. The underlying infrastructure is based on a complex relationship model among three core entities: people, items, and tags. We describe the general model and the different recommender systems that were built on top, including the main results and the implications from one system to another. We conclude by highlighting the main findings and suggesting next steps and future directions.

6.1 Introduction

Enterprise social media has emerged as a mean to share and interact within the enterprise. As on the web, employees have started to use social media tools behind the firewall, such as social bookmarking [17], corporate blogging [15], corporate wikis [3], enterprise file sharing [18], and enterprise social network sites [4]. In addition, types of applications that are more unique to an enterprise have emerged, taking advantage of the single identity behind the firewall and the high accountability employees assume to it. Examples for such applications include people tagging [6] and human computation games [10].

Similarly to social media outside the firewall, the prosperity of enterprise social media leads to one of its greatest challenges –information overload. In fact, in the context of social media, the overload is not of information only, but also of interaction: the number of friends, followers, followees, co-members, commenters, taggers, "likers", and so forth often leads to an overdose of interaction. *Social recommender systems* (SRSs) [8, 9] aim to alleviate the combination of information and interaction overload, to which we refer as *social overload*. Often applying personalization

Ido Guy
IBM Research-Haifa, 31905, Israel
e-mail: ido@il.ibm.com

J.J. Pazos Arias et al.: Recommender Systems for the Social Web, ISRL 32, pp. 121–135.
springerlink.com

techniques, SRSs try to filter information and interaction by adapting those to the need of the individual user.

In an enterprise setting, social overload is reflected in employees being exposed to large amount of content, including blogs and microblogs, wikis, communities, files, or bookmarks, and also to other employees who want to connect or follow. Such enterprise social overload is obviously more acute in large organizations with tens of thousands and sometimes hundreds of thousands of employees, often spread worldwide. In order to accurately personalize the recommendations, a rich user model needs to be built. In addition, a recommender system needs to deal with the cold start problem [20], both of new items and of new employees. A rich aggregation model, which can draw information from multiple organizational systems, from the corporate directory and the organizational chart, through project and publication databases, to blogging, wikis, and file sharing systems, can help address these challenges. In an enterprise environment, aggregation becomes more feasible and effective due to the fact that employees typically have a single identity in all enterprise systems.

In this chapter, we describe our enterprise aggregation system, and then a set of social recommender systems, for both people and content, which leverage this aggregation system. We describe a rich set of results within an enterprise scope and implications for future work.

The rest of this chapter is organized as follows. The next section shortly describes our aggregation model. The following section describes two people recommender systems (RS) we built over the model: the first for familiar people and the second for strangers. We then describe two studies on recommendation of social media content: the first using social relationships only and the second using tags as well. We conclude by discussing our results and suggesting a few directions future work should explore.

6.2 Social Agregation Model

SaND [19] is an aggregation system that models relations between people, resources (referred to as "items" in the remainder of this chapter), and tags through data collected across the enterprise, and in particular across enterprise social media applications. SaND aggregates any kind of relationships between its three core entities –people, items, and tags. The implementation of SaND is based on a unified approach [1], in which all entities are searchable and retrievable. As part of its analysis, SaND builds an entity-entity relationship matrix that maps a given entity to all related entities, weighted according to their relationship strength. The entity-entity relationship strength is composed of two types of relations:

- *Direct Relations*: Figure 6.1 shows all direct relations between entities that are modeled by SaND. Particularly, a user is directly related to: (i) another person: as an explicit friend, as a tagger of or tagged by that person, or through the organizational chart (direct manager or employee); (ii) an item (e.g., a bookmarked web page, a shared file, or a community): as an author, a commenter, a tagger,

or a member; (iii) a tag: when used by the user or when applied on the user by others. In addition, an item is directly related to a tag if it was tagged with it.
- *Indirect Relations*: two entities are indirectly related if both are directly related to another entity. For example, two users are indirectly related if both are related to another user, e.g., if both have the same manager or friend or if both have tagged or were tagged by the same person.

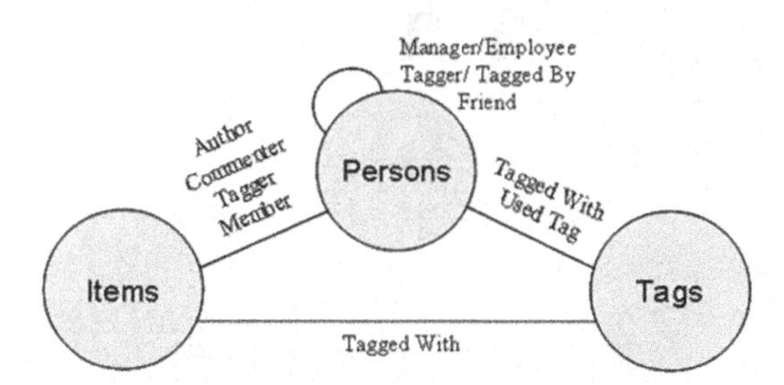

Fig. 6.1 SaND Relation Diagram

For each type of direct or indirect relationship, SaND applies a relationship weight. For example, an authorship person-item relation is stronger than a tagging relation; the weight of a co-authorship person-person relationship is affected by the number of other co-authors. The ultimate relationship weight between two entities is determined by summing over the weights of all direct and indirect relationships between them.

For people-to-people relationships, SaND distinguishes between familiarity relationships (people the user knows) and similarity relationships (people whose social activity is found to be overlapping with the user's social activity). Familiarity relationships include all direct person-person relations, as well as two types of indirect relations: co-authorship (e.g., of a file or a wiki) and having the same manager. Similarity relationships include indirect relations only, such as co-usage of the same tag, co-tagging of the same item, co-commenting on the same blog entry, or co-membership in the same community.

6.3 People Recommendation

6.3.1 Recommending Familiar People

With the proliferation of social networks sites (SNSs) [2], allowing users to connect with each other by sending and accepting invitations from one another, the need for effective people recommendation systems has become evident. This topic strongly relates to the area of social matching [21], which discusses RS that recommend people to people and their uniqueness in aspects such as privacy, trust, reputation, or interpersonal attraction. In parallel to the emergence of "people you may know" widgets on leading SNSs, such as Facebook and LinkedIn, the work on the "do you know?" (DYK) widget [11] was the first to study the topic. The widget recommends people to connect with in the enterprise, based on a rich set of implicit people-to-people relationships, derived by SaND.

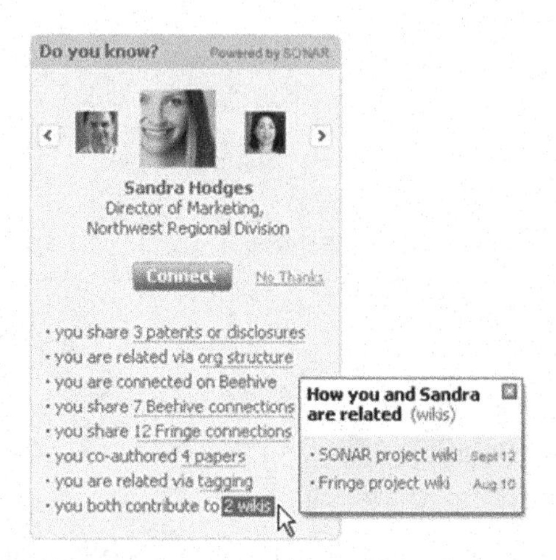

Fig. 6.2 The "Do you know?" widget for people recommendation

Figure 6.2 illustrates the DYK widget. It enables the user to scroll though a list of recommended people one at a time. The list of people is retrieved by requesting the top 100 related people to the user as retrieved by SaND. The recommended people are presented in descending order of relationship score. Hence, the first recommendations are those with whom the user has the strongest familiarity level and is not connected to yet. For each recommendation, the widget presents a picture, the person's name as a link to her profile, and a summarized list of all available evidences of the relationship to the user. These evidences are retrieved from SaND as well. The summary would state, for example, that the recommended person and the user wrote two papers together, commented on each other's blogs three times, tagged each other in the people tagging application, share a manager, have 10 mutual connections, and so forth. While hovering over each summarized evidence item, a popup appears which includes a detailed list of evidences with relevant links. In Figure 6.2, the popup shows the two wikis the users share. After viewing a recommended person, the user can decide to scroll to the next or previous person, to invite the current person to connect, or to remove the current recommendation. The user invites the recommended person by clicking on a connect button below the name of the person. This will bring up a default invitation text which users can edit according to their wishes. The invitation is then sent out to the recommended person through email and the DYK proceeds to the next recommended person.

The DYK widget was deployed as part of the homepage of a widely used enterprise SNS, called Fringe [6]. A field study, inspecting the widget's usage over four months, reported a dramatic effect on the site, both in terms of number of invitations sent and in terms of number of people who send invitations [11].

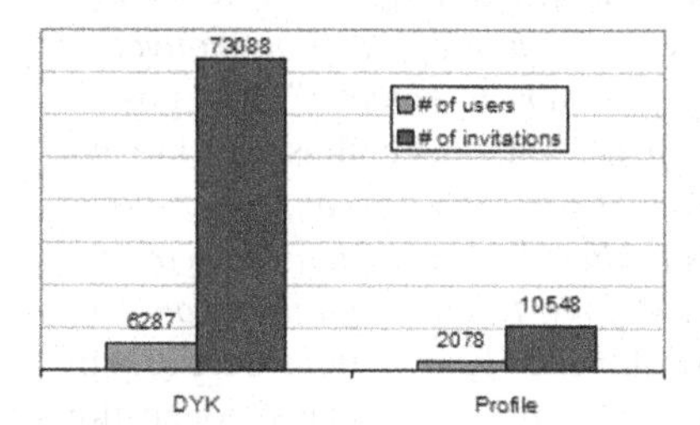

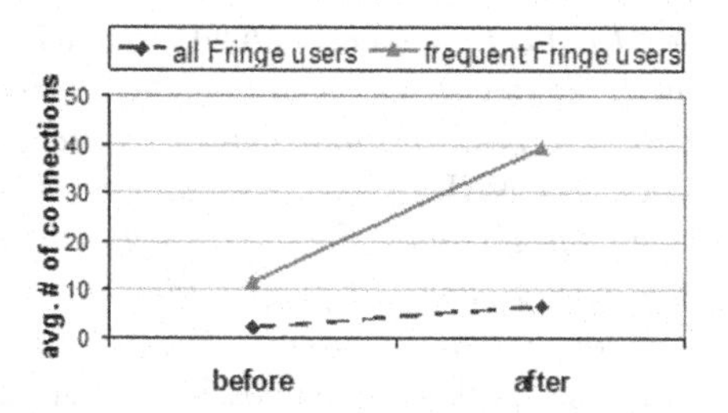

Fig. 6.3 Effects of the DYK widget on the Fringe Site. On the left, the number of invitations sent and the number of user sending invitations through the DYK widget, compared to the regaulr profile mechaisnm. On the right, the increase in number of connections for frequent and regular users

Figure 6.3 illustrates these results. The diagram on the left compares the usage of the DYK widget with the usage of the regular mechanism of inviting through people profiles during the four month period. First, it shows the number of invitations sent from the DYK widget compared to the number of invitations sent from others' profiles –73,088 invitations were sent through the DYK widget, while only 10,548 were sent through profiles. Acceptance rate was exactly 60% both for invitations sent from DYK and for invitations sent from profiles. The identical acceptance rate indicates that while the DYK widget provoked much more invitations, their quality in terms of acceptance rate remained equal to that of the usual profile-based mechanism. In addition, while 6287 users initiated invitations though the DYK widget, only 2048 sent invitations through profiles. These high differences between DYK invitations and profile invitations stand in contrast to the fact that the homepage was accessed 79,108 times, while profile pages were accessed 91,964 times. The overall increase in the number of invitations in Fringe was 278%, which resulted in an overall increase of 230% in confirmed connections. The increase in people who sent at least one invitation was 150%. We note that this sharp increase took place within a period of four months and after the Fringe "friending" feature had been available for 15 months without recommendations.

The diagram on the right of Figure 6.3 shows the substantial change in the average number of connections per user after the DYK was introduced. Frequent users, who accessed Fringe for at least 10 days during the four-month period, had 11.6 connections on average before the DYK (stdev 22.7, median 4, max 198), and ended up with 39.1 connections on average (stdev 37.2, median 30, max 389). All Fringe users together had 2.0 on average at the beginning of the period (stdev 6.3, median 1, max 198) and 6.6 on average after the period (stdev 12.2, median 2, max 389).

Out of the 6287 individuals who used the DYK widget to invite other colleagues, 5076 (80.7%) had sent no invitations at all before using the DYK. This high percentage implies that the easiness of inviting others through the DYK widget relative to traditional methods broke the entry barrier for many users and caused them to start playing the "social game" of connecting to others. This assumption was reinforced in blog posts that referred to the DYK widget: one blogger wrote *"I must say I am a lazy social networker, but Fringe was the first application motivating me to go ahead and send out some invitations to others to connect"*. In a post titled *"Fringe... I'm out of control! Somebody stop me!"* another enthusiastic blogger said *"I've **NEVER** seen such an easy way to invite someone. I mean, that rollover thingie to invite people to connect with you is addictive. In a matter of seconds, I sent invitations to 28 people. Me! The oh, I'm choosy, I don't send to anyone and everyone social networker"*. The fact that many of the DYK invitations were sent by "newbies", rules out the possibility that recommendations only inflated connection lists of users who had connections before.

Another noticeable aspect of the DYK was the relatively high decay of usage along time. Most users used it only in a few sessions (typically between 1 to 10 sessions), until they have exhausted the recommendations and built a rich enough network. While enterprise relations continue to form over time, clearly the need for friend recommendation is most acute in early stages of network building. In addition, users of the DYK widget also expressed their desire to get more "interesting" people recommendation, of employees they do not yet know. For example, one user commented: *"at the moment, it keeps you limited to whom you already know, but can be useful to recommend similar people to extend your reach"*. This led us to explore a second type of people recommendation in the enterprise.

6.3.2 Recommending Strangers

Research on SNSs within the enterprise has indicated that in addition to staying in touch with close colleagues, employees use enterprise SNSs to reach out to employees they do not know and build stronger bonds with their weak ties. Their motivations include connecting on a personal level with more coworkers, advancing their career within the company, and campaigning for their ideas [4]. Our StrangerRec system [12] recommends employees the user is not familiar with, but may be of interest based on common behavior on enterprise social media, such as usage of the same tags or commenting on the same blogs.

The task of recommending unfamiliar yet interesting people is quite different from "regular" recommendation of familiar people. Our recommender focuses more on discovery and exposure to new people and less on facilitating connection within an SNS. It aims at satisfying two rather conflicting goals: on the one hand, the recommended person should not be familiar to the user, and, on the other hand, the person should be of some interest. While accuracy of recommendations that satisfy both goals might not be high, we argue that the potential serendipity and "surprise effect" in getting a fortuitous recommendation of an interesting new person in the organization may compensate for lower accuracy [16].

To implement our stranger recommender, we used the people-to-people relationship modeling of SaND, and leveraged SaND's capability to distinguish between two types of people relationships: familiarity and similarity. Familiarity relationships are derived based on indicators for knowing a person: either explicit indicators (being connected on a SNS or a connection through the organizational chart) or implicit indicators (co-authoring a wiki page, being member in the same project, sharing a file, etc.). Similarity relationships are derived based on common activity in social media, which serves as an implicit indicator for mutual interests. For example, usage of the same tag, being tagged by the same tag within the people tagging application, bookmarking the same page, membership in the same community, or commenting on the same blog, are all considered similarity relationships. To generate the recommendations, we apply a *social network composition*, i.e., a composition of two social network types. In tis specific case, we subtract the user's familiarity network (i.e., list of people s/he knows) from the similarity network (list of people s/he has common interests with) to suggest strangers who may be of interest. The rich underlying aggregation model ensures that we can derive many types of similarity relationships, while also being able to effectively filter out people the uer is already familiar with.

Figure 6.4 demonstrates the user interface of StrangerRS. Part A shows the profile page of the recommended employee. As opposed to the DYK widget, where only few details (photo, name, job title) were shown, here the entire profile was exposed in order to provide as many hints on the recommended stranger as possible (e.g., their office location, management chain, friends, tags applied by others, or board messages). Part B shows the evidence for similar interests with the recommended individual, for instance, communities they are both member of, tags they have both used, or blogs they have both commented on. Part C shows the feedback users were asked to provide on the recommendations. Particularly, the first question asks whether the user knows the recommended person and the second asks whether the recommended person is of interest.

Figure 6.5 summarizes the rating results. The upper part refers to rating of Q1. We compared StragnerRS with two benchmarks: a random person (Random) and a strongly familiar person (StrongFam). It can be seen that over two thirds (67.3%) of StrangerRS recommendations are indeed strangers, compared to 97.7% of the random recommendations and only 4.6% of the strongly familiar recommendations. Hence, StrangerRS is able to recommend people who are likely to be strangers, even if not with the same likelihood as a random person.

The bottom part of Figure 6.5 shows the rating results of Q2 given that Q1 was rated 1 (i.e., the person was a stranger). While 40% of these recommendation by StrangerRS were rated as 1 (non-interesting), nearly 60% raised some interest (compared to 30% for random), and 30% were rated 3 or more. Overall, even though Stranger RS' likelihood to recommend a stranger is lower than Random, its likelihood to recommend an **interesting** stranger is higher. The latter statement is obviously true in comparison to the StrongFam. Out of 9 recommendations, StrangerRS

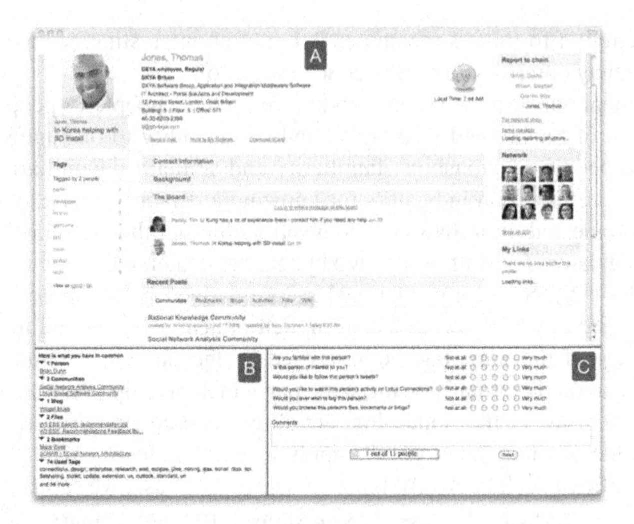

Fig. 6.4 User interface of the stranger recommender systems

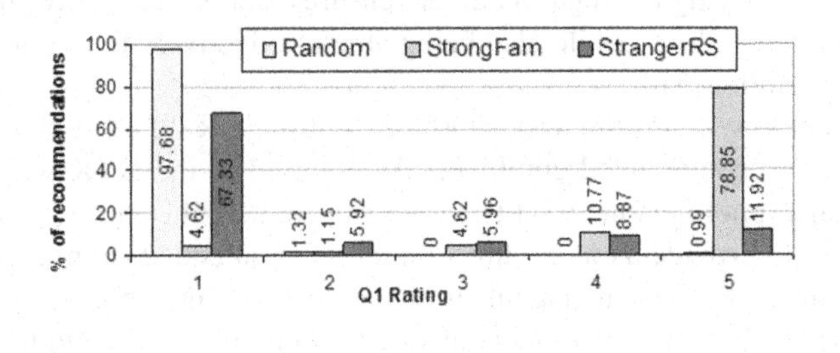

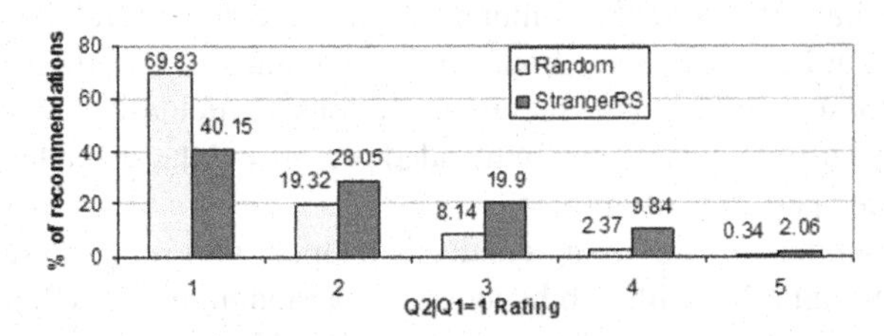

Fig. 6.5 Rating results for StrangerRS compared to the Random and StrongFam benchmarks. The upper part shows the ratings of Q1 (familiarity with the recommended person). The bottom part shows the ratings of Q2 (interest in the recommended person) when Q1=1 (the person is a stranger)

was able to recommend at least one stranger rated with Q2=3 or higher for over two thirds of the users, and at least one stranger with Q2=4 or higher for over 36% of the users. The value of such a recommendation, suggesting a stranger who is interesting, can be very high to workers. We received several reasons for interest in strangers within the organization. For example, one user said: *"Following this person might help me better understand the sales environment in his part of the world"* and another mentioned: *"She works for a marquee customer in the Telecom sector I cover. Any lessons or best practices she shares I would be very much interested in"*. A third one commented: *"Works with implementations of products I work with. Other key contacts are known by this person. Useful tags. Looks useful.*

In conclusion, while stranger recommendations have significant lower accuracy than recommendations of known people, their value lies in other aspects, such as serendipity and diversity. Practically, along time, SNSs should combine both types of recommendations. For example, friend recommendations can be suggested to new social media users who are building their initial network. Once established, stranger recommendations can help extend social circles and expand reach. Another option is to mix both friend and stranger recommendations in parallel, integrating both the higher accuracy of friend recommendations and the serendipity, or "surprise effect", of stranger recommendations. Further research needs to examine in detail how to interleave both types of recommendations.

6.4 Content Recommendation

6.4.1 Recommendation Based on Social Relationships

As enterprise social media becomes more popular, employees are exposed to a variety of social media applications behind the firewall, such as social bookmarking systems [17], corporate blogs [15], microblogs [5], wikis [3], enterprise SNSs [4, 6], enterprise file sharing [18], and more. We next describe our experimentation with recommending mixed social media items to employees. The platform we experimented with was Lotus Connections (LC)[1], an enterprise social software application suite that includes different social media applications: bookmarks, blogs, wikis, communities, and files, among others. LC also includes a rich corporate directory with employee profiles, allowing employees to connect to each other as in an SNS, tag each other, and write status updates ("microblogs").

We initially focused on recommending three types of items –bookmarks, blogs, and communities– based on people relationships only [13]. To this end, we again leveraged the people-to-people relationship model within SaND and the distinction between familiarity and similarity. Our focus was on two main research questions:

1. Which of the two relationship types is more effective for recommendation of social media items? On the one hand, people who share similar activity with you, not necessarily your friends, are the ones who are likely to indicate the most attrative items for you. On the other hand, in real life people mostly seek advice from people they know. Within an enterprise, the people you are familiar with are typically colleagues with whom you work or have worked in the past, and are thus likely to be a source for interesting items.

[1] IBM Social Software for Business- Lotus Connections:
`http://www.ibm.com/software/lotus/products/connections`

2. Can explanations, detailing the people who yield the recommendation and their relationships to the user and the recommended item, improve the quality of recommendations?

The two questions relate to each other to some extent, since explanations can affect the comparison between the familiarity and similarity networks. In essence, explanations are likely to be more effective for the familiarity network, as the user can be affected by the fact that a familiar person is related to a recommended item and may also be able to better judge the recommendation once it is associated with a familiar person.

Figure 6.6 depicts the item recommendation widget. The user is presented with five items consisting of a mix of bookmarked pages, communities and blog entries. Each item has a title which is a link to the original document and a short description if available. The icon to the left of each item represents its type –the first item in Figure 6.6 is a blog entry, the second is a community, and the fourth is a bookmarked page. The user can remove an item in order to retrieve a new recommendation by clicking on the Next icon. Each recommended item includes a list of person names that are related to the item. Each person provides an explanation of why the item is recommended (serving as an implicit recommender of the item). When hovering over a name, the user is presented with a popup detailing the relationships of that person to the user and to the item. In Figure 6.6 the recommended items are chosen according to the similarity network of the user. The popup indicates that Ido on the one hand is a member of the recommended community and on the other hand is similar to the user as they both share a set of documents and used the same tags.

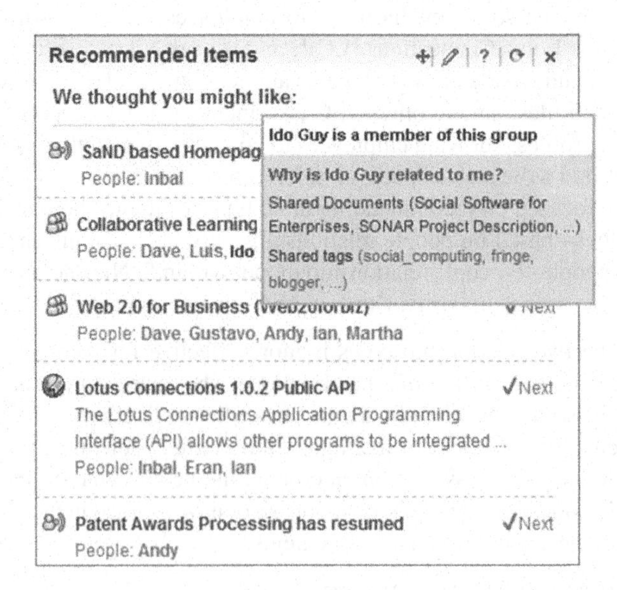

Fig. 6.6 Widget for recommending social media items based on social relationships

The recommender engine recommends items according to the following formula (representing the score of item i for user u):

$$RS(u,i) = e^{-\alpha \cdot t(i)} \cdot \sum_{v \in N^T(u)} S^T[u,v] \sum_{r \in R(v,i)} W(r)$$

where $t(i)$ is the number of days passed since the creation date of i; α is a decay factor (set in our experiments to 0.025); $N^T(u)$ is the set of users within u's network of type T; $S^T[u,v]$ is the SaND relationship score between u and v based on the network of type T; $R(v,i)$ is the set of all relationship types between user v and item i, given by SaND (authorship, membership, etc.); and $W(r)$ is the corresponding weight for the user-item relationship type r as given by SaND.

Our main evaluation of the above RS was based on a user study with 290 participants who were evenly assigned to one of three groups: familiarity, similarity, and overall (the latter is a combination of both types of networks). Each participant rated 12 recommended items in two randomly-ordered phases: one phase included explanations for each recommended item, while the other did not include explanations.

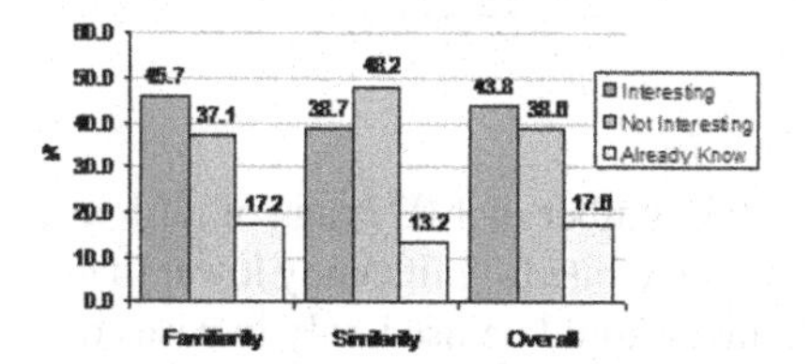

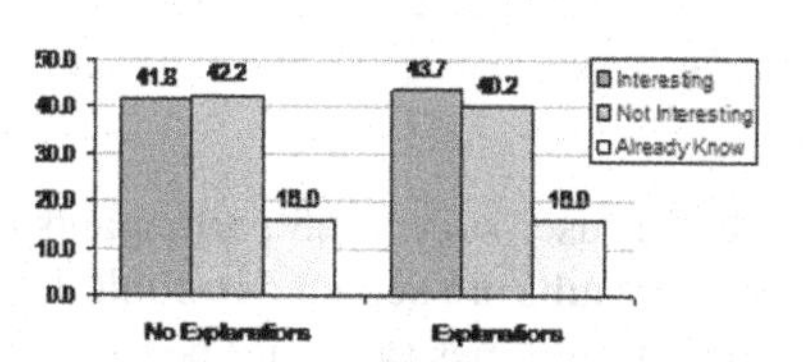

Fig. 6.7 Rating Results for item recommender based on social relationships. On the left, familiarity-based recommendations are rated more interesting than similarity-based recommendations. On the right, explanations slightly increase the interest rate in recommended items

Figure 6.7 shows the main results of the survey. On the left, the comparison between the three relationship types: familiarity, similarity, and overall. Familiarity-based recommendations are significantly more interesting than similarity-based recommendations, showing that in terms of accuracy, familiarity yields better recommendations. The similarity network, however, has lower percentage of already-known items, indicating that in terms of serendipity and novelty, this network may have an advantage (at the expense of accuracy). Combining the two networks increases accuracy, but also expectedness (already know rate is similar to the familiarity network).

The right side of Figure 6.7 indicates the effect of explanations. Recommendations with explanations were rated slightly higher than recommendations with no explanations. This finding was not statistically significant, but was consistent across all three network types and most noticeable for the familiarity network. It indicates that in addition to longer-term benefits, people-based explanations also instantly affect the interest in recommended items.

6.4.2 Recommending Based on Tags

After exploring people-based recommendations, we moved on to examine the effect of adding tag-based recommendations. We first examined several types of tags as indicators of users' interests. We examined tags used by the user in the different tagging systems indexed by SaND ("used tags"), tags applied on the user by others through the people tagging application ("incoming tags"), a combination of used and incoming tags ("direct tags"), and tags applied on items the user is related to ("indirect tags").

Table 6.1 Rating results of tags as topics of interest

%	Not interested	Interested	Highly interested
Used	16.84	38.25	44.91
incoming	15.48	31.75	52.78
direct	7.46	22.81	69.74
indirect	35.38	45.38	19.23

In our user survey, 65 participants rated a total number of 1,037 tags of all four types. Results are depicted in Table 1. Indirect tags are rated significantly lower than all other tag types (as could be expected; such tags should be used only in cases of data sparsity). Direct tags are rated significantly higher than all other types of tags, indicating that tags that are both used by the user and applied to her by others are the most effective interest indicators. Interestingly, incoming tags are rated slightly higher than used tags (insignificant difference), indicating that the topics associated with the user by the crowd are as good indicator for the user's topics of interests as the tags she used herself.

Based on these results, we built a hybrid people-tags-based recommender that suggests five types of social media items: bookmarks, blogs, communities, files, and wikis [14]. The recommender is based on a user profile that consists of both people and tags. For people, familiarity relationships are favored over similarity relationships by a factor of 3, due to the results of the people-based recommender. For tags, we considered direct tags, due to the results of the tags user survey described above. Overall, items are recommended according to the following formula:

$$RS(u,i) = e^{-\alpha \cdot d(i)} \cdot \{\beta \cdot \sum_{v \in N(u)} w(u,v) \cdot w(v,i) + (1-\beta) \cdot \sum_{t \in T(u)} w(u,t) \cdot w(t,i)\}$$

where $d(i)$ is the number of days passed since the creation date of i; α is a decay factor (set in our experiments to 0.025); β is a parameter that controls the relative weight between people and tags and is used in our experiments to evaluate different recommenders; $w(u,v)$ and $w(u,t)$ are the relationship strengths of u to user v and tag t, as given by the user profile; $w(v,i)$ and $w(t,i)$ are the relationship strengths between v and t, respectively, to item i as determined by the direct relations in SaND.

Ultimately, the recommendation score of an item, reflecting its likelihood to be recommended to the user, may increase due to the following factors: more people and/or tags within the user's profile are related to the item; stronger relationships of these people and/or tags to the user; stronger relationships of these people and/or tags to the item; and freshness of the item. We exclude items that are found to be directly related to the user. For example, we will not recommend an item the user has already commented on or has already tagged.

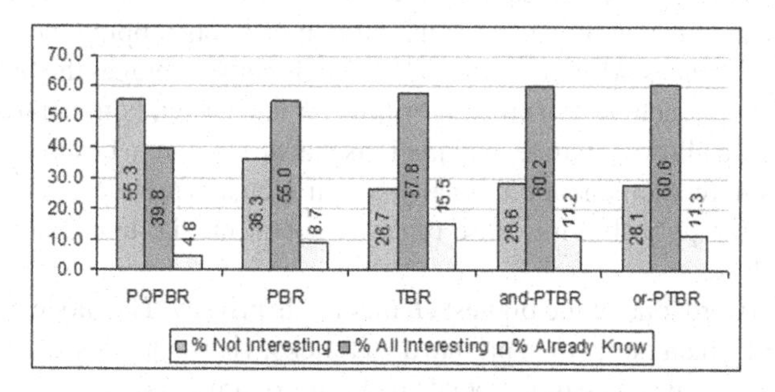

Fig. 6.8 Item Rating Results across 5 recommenders: a popularity-based recommender, a recommender based on people only, a recommender based on tgas only, and two hybrid recommenders combining both people and tags

In our main user survey, we compared among 5 types of recommenders: (i) POPBR –a non-personalized popularity-based recommender, as a benchmark, (ii) PBR –a people-based recommender, similar to the one used in the first study, considering social relationships only, (iii) TBR –a tag-based recommender considering tags only, (iv) and-PTBR –a hybrid recommender based on both people and tags, (v) or-PTBR –a hybrid recommender based on people or tags.

Results are depicted in Figure 6.8. All types of personalized recommenders significantly outperform the popularity-based recommender. The tag-based recommender outperforms the people-based recommender in terms of accuracy (interest ratio in recommended items). However, it poses a few shortcomings compared to the people-based recommender: higher level of expectedness, reflected in a higher percentage of already-known items; lower level of diversity, reflected in about 80% of the recommended items being bookmarked pages; and lower effectiveness of explanations. Both hybrid recommenders enjoy the benefits of both "worlds" –they maintain as high accuracy as the tag-based recommender, and substantially improve on diversity and serendipity.

6.5 Conclusion and Future Work

We presented a rich set of recommender systems that suggest both people and content within the enterprise. All RS rely on the SaND system, which aggregates relations among people, items, and tags across various enterprise systems. The fact that employees have single identities facilitates the aggregation task. The rich aggregation model allows generating high-quality recommendations that reach high level of accuracy, based on a diverse and comprehensive set of signals. It also allows dealing with the cold start problem of new users by relying on external information (see [7] for more detail).

Our systems rely on user implicit interest indicators, such as the user's social relationships and tags, and thus spare the burden of bootstrapping the system with initial user preferences. Moreover, the fact that it directly leverages the SaND model, without applying machine-learning algorithms or clustering, both contributes to its performance and allows intuitive explanations that are shown to be effective for both people and item recommendation. Finally, as our methods rely on a general relation model, they are applicable for more types of social media items, such as music, photos, or videos.

Moving forward, one of the biggest challenges is privacy. The model we describe leverages public data rather than private data (clicks, views, queries, etc.) and is thus less sensitive to privacy issues. Social media users are exposed to privacy issues when presented with recommendations, especially when those are accompanied by explanations (e.g., "because you have watched this video"), due to the fact that other users may access their account or view their screens. Future social recommender systems should put special care on privacy preservation.

In the analysis above, we discussed factors other than accuracy, such as diversity and serendipity. Yet, the focus of the evaluation is primarily on accuracy and no predefined methodology is used to evaluate the combination of different evaluation measures. More emphasis on developing new evaluation methods, with focus on factors other than accuracy, should be put in future social recommendation research. Moreover, evaluation over time, where users get used to the system and exhaust the initial set of recommendations, while their interests keep changing, is also due for future studies. Learning from user feedback can help mitigate the expected decrease of user interest in recommendations along time.

The focus of all of our works has been on enterprise systems. Some of our results have implications for social media sites outside the firewall, due to the fact that most of the systems we inspected have analogous versions outside the firewall. Yet, applying a similar aggregation model outside the firewall is one of the biggest challenges ahead, and requires handling multiple identities and much higher volumes of data and users.

Acknowledgements. With thanks to David Carmel, Tal Daniel, Michal Jacovi, Shila Ofek-Koifman, Adam Perer, Inbal Ronen, Sigalit Ur, Erel Uziel, Eric Wilcox, Sivan Yogev, and Naama Zwerdling for jointly working on the studies described in this chapter.

References

1. Amitay, E., Carmel, D., Harel, N., Soffer, A., Golbandi, N., Ofek-Koifman, S., Yogev, S.: Social search and discovery using a unified approach. In: Proc. HYPERTEXT 2009, pp. 199–208 (2009)
2. Boyd, D.M., Ellison, N.B.: Social network sites: Definition, history, and scholarship. Journal of CMC 13, 1 (2007)
3. Danis, C., Singer, D.: A wiki instance in the enterprise: opportunities, concerns and reality. In: Proc. CSCW 2008, pp. 495–504 (2008)
4. DiMicco, J., Millen, D.R., Geyer, W., Dugan, C., Brownholtz, B., Muller, M.: Motivations for social networking at work. In: Proc. CSCW 2008, pp. 711–720 (2008)
5. Ehrlich, K., Shami, N.S.: Microblogging inside and outside the workplace. In: Proc. ICWSM 2010 (2010)
6. Farrell, S., Lau, T., Nusser, S., Wilcox, E., Muller, M.: Socially augmenting employee profiles with people-tagging. In: Proc. UIST 2007, pp. 91–100 (2007)
7. Freyne, J., Jacovi, M., Guy, I., Geyer, W.: Increasing engagement through early recommender intervention. In: Proc RecSys 2009, pp. 85–92 (2009)
8. Guy, I., Carmel, D.: Social Recommender Systems. In: Proc. WWW 2011, pp. 283–284 (2011)
9. Guy, I., Chen, L., Zhou, M.X.: Workshop on social recommender systems. In: Proc. IUI 2010, pp. 433–434 (2010)
10. Guy, I., Perer, A., Daniel, T., Greenshpan, O., Turbahn, I.: Guess who? enriching the social graph through a crowdsourcing game. In: Proc. CHI 2011 (2011)
11. Guy, I., Ronen, I., Wilcox, E.: Do you know? recommending people to invite into your social network. In: Proc. IUI 2009, pp. 77–86 (2009)
12. Guy, I., Ur, S., Ronen, I., Perer, A., Jacovi, M.: Do you want to know? recommending strangers in the enterprise. In: Proc. CSCW 2011, pp. 285–294 (2011)
13. Guy, I., Zwerdling, N., Carmel, D., Ronen, I., Uziel, E., Yogev, S., Ofek-Koifman, S.: Personalized recommendation of social software items based on social relations. In: Proc. RecSys 2009, pp. 53–60 (2009)
14. Guy, I., Zwerdling, N., Ronen, I., Carmel, D., Uziel, E.: Social media recommendation based on people and tags. In: Proc. SIGIR 2010, pp. 194–201 (2010)
15. Huh, J., Jones, L., Erickson, T., Kellogg, W.A., Bellamy, R.K., Thomas, J.C.: BlogCentral: the role of internal blogs at work. In: Proc. CHI 2007, pp. 2447–2452 (2007)
16. McNee, S.M., Riedl, J., Konstan, J.A.: Being accurate is not enough: how accuracy metrics have hurt recommender systems. In: Proc. CHI 2006, EA, pp. 1097–1101 (2006)
17. Millen, D.R., Feinberg, J., Kerr, B.: Dogear: social bookmarking in the enterprise. In: Proc. CHI 2006, pp. 111–120 (2006)
18. Muller, M., Millen, D.R., Feinberg, J.: Patterns of usage in an enterprise file-sharing service: publicizing, discovering, and telling the news. In: Proc. CHI 2010, pp. 763–766 (2010)
19. Ronen, I., Shahar, E., Ur, S., Uziel, E., Yogev, S., Zwerdling, N., Carmel, D., Guy, I., Harel, N., Ofek-Koifman, S.: Social networks and discovery in the enterprise (SaND). In: Proc. SIGIR 2009, p. 836 (2009)
20. Schein, A.I., Popescul, A., Ungar, L.H., Pennock, D.M.: Methods and metrics for cold-start recommendations. In: Proc. SIGIR 2002, pp. 253–260 (2002)
21. Terveen, L., McDonald, D.W.: Social matching: a framework and research agenda. ACM Trans. Comput.-Hum. Interact. 12(3), 401–434 (2007)

Part IV
Group Recommendation

Chapter 7
Group Recommender Systems: New Perspectives in the Social Web

Iván Cantador and Pablo Castells

Abstract. An increasingly important type of recommender systems comprises those that generate suggestions for groups rather than for individuals. In this chapter, we revise state of the art approaches on group formation, modelling and recommendation, and present challenging problems to be included in the group recommender system research agenda in the context of the Social Web.

7.1 Introduction

Social Web technologies have emerged as a new step in the course of technological innovations that are having an impact on our everyday lives, reaching the way people relate to each other, work, learn, travel, buy and sell, discover new things, make themselves known, or spend their leisure time. From the common user perspective, while prior technological breakthroughs have to a large extent empowered the individual (giving her instant access to universal online information and services, public authoring access to worldwide publication channels, portable network endpoints, audiovisual production devices, custom-fit adaptation of services to the individual user, etc.), the new trend explicitly emphasises social awareness. As is studied e.g. in Social Sciences, society as a human phenomenon comes along with the notion of group. Groups are indeed a cardinal element in all spheres of social interaction and behaviour. Be they organisations, clubs, political parties, family, tribes, professional units, circles of friends, or just occasional gatherings of people, groups have a part in most human activities, and have played a central role in the evolution of mankind across the ages.

Iván Cantador · Pablo Castells
Universidad Autónoma de Madrid
28049 Madrid, Spain
e-mail: {ivan.cantador,pablo.castells}@uam.es

J.J. Pazos Arias et al.: Recommender Systems for the Social Web, ISRL 32, pp. 139–157.
springerlink.com

The new social environments open up new possibilities to define, form, articulate, manage and leverage group structures for multiple purposes. The available infrastructure, the explosive growth of online communities, their increasing activity and collected data, lift boundaries and multiply the possibilities to model groups as complex units and draw added value from them. Different degrees of group existence can be considered, from sets of people that meet, interact, or have some actual common bond in the physical world, to online contacts that have no relation outside the system, to latent groups of users that are not even directly aware of each other. The new perspectives bring an opportunity to creatively conceive new views and roles for groups in social environments, and perhaps a new angle on the traditional tension between the individual and the group.

As a particular case, in this chapter, we focus on the role of groups in recommender systems. Recommendation technologies are one of the most successful areas of ongoing innovation which find a natural environment to play their best on the Social Web, given the wealth of user input, multiple evidence of user interest, and the huge scale of the new social spaces, where users often count by the million – or billion. Recommender systems have traditionally targeted individual users as the recipients of the personalised system's output. The perspective of delivering shared recommendations for groups as a whole is a new take on the recommendation task that has recently started to be addressed in the field.

The motivation and usefulness of group recommendations naturally arises in situations where a group of users shares a common activity, service, task, or device. For instance, a recommender system could suggest a movie or TV show to watch by a particular group of people (a couple, a family, a group of friends), or could select a sequence of music tracks to be played in a place where individuals with multiple tastes cohabite (a bar, a gym, a shop). Group-oriented recommendation is useful as well in commonplace scenarios such as planning a trip, or choosing a restaurant. Also, in general, many scenarios where ambient intelligence (a.k.a. pervasive/ubiquous computing) technologies take place, and where different people cohabit for a period of time, are susceptible to incorporate group recommendation functionalities. There is a wide number of works in the research literature addressing the group modelling and recommendation problems [16, 17], in different domains and applications, such as recommending tourist attractions [2, 15, 20, 18], food recipes [5], TV programs and movies [4, 10, 23, 27, 36], video clips [18], music tracks and radio stations [8, 21], photos [7], and Web and news pages [25, 29], to name a few.

The group-based perspective changes the recommendation task not only in its purpose, but also in the starting conditions. For instance, the decision of a group member whether or not to accept a given recommendation may depend not only on her own evaluation of the content of the recommendation, but also on her beliefs about the evaluations of the other group members, and about their motivation. As pointed out by Masthoff in [19], the opinion of other members of the group may influence the opinion expressed by a particular user, based on the so-called process of *conformity*, while, on the other hand, the satisfaction of other group members can also lead to increase the user's satisfaction by the so-called *emotional cognition process*. The groups may be quite heterogeneous, in terms of age, gender, intelligence and personality influence on the perception and complacency with the system outputs each member of the groups may have. Thus, a major question that arises is how a recommender system can adapt itself to a group of users, in such a way that each individual enjoys and benefits from the results.

Nowadays, in Web 2.0 systems, people communicate online with contacts through social networks, upload and share multimedia contents, maintain personal bookmarks and blogs, post comments and reviews, rate and tag resources available on the Web, and contribute to wiki-style knowledge bases. The huge amount of user generated content, together with the complexities and dynamics of large groups of people in the Social Web provide room for further research on group recommender systems.

In this chapter, we revise state of the art approaches to group formation, modelling and recommendation, and present challenging problems to be included in the group recommender system research agenda in the context of the Social Web. The rest of the chapter is organised as follows. In Section 2, we explain strategies based on Social Choice Theory, which have been taken into account by the existing group recommendation approaches. In Section 3, we describe group recommender systems presented in the literature, covering different aspects such as group formation, group profile modelling, recommendation aggregation, and cooperative consensus. In Section 4, we revise several open research lines in group recommendation, and in Section 5, we discuss additional challenges that arise in group recommender systems for the Social Web. Finally, in Section 6, we end with some conclusions.

7.2 Social Choice Theory

Though recommendation approaches have addressed group preference modelling explicitly to a rather limited extent, or in an indirect way in prior work in the computing field, the related issue of *social choice* (also called *group decision making*, i.e. deciding what is best for a group given the opinions of individuals) has been studied extensively in Economics, Politics, Sociology, and Mathematics [24, 33]. The models for the construction of a social welfare function in these works are similar to the group modelling problem we put forward here.

Other areas in which Social Choice Theory has been studied are Collaborative Filtering (CF), Meta-search, and Multi-agent systems. In CF, preferences of a group of individuals are aggregated to produce a predicted preference for somebody outside the group. Meta-search can be seen –and formulated– as a form of group decision making, where the aggregated inputs are produced by information retrieval systems instead of people. In a meta-search engine, the rankings produced by multiple search engines need to be combined into one single list, forming the well-known problem of *rank aggregation* in Information Retrieval [3]. Ensemble recommenders combining several recommendation algorithms also involve a particular case of this problem, similarly to meta-search except for the absence of a query. Finally, in

Multi-agent systems, agents need to take decisions that are not only rational from an individual's point of view, but also from a social point of view.

In all the above fields, different strategies to combine several users' preferences and to aggregate item ranking lists can be applied based on the utilised social welfare function. These strategies are classified by Senot and colleagues [27] into three categories, namely *majority-based strategies*, which strength the "most popular" choices (user preferences, item rankings, etc.) among the group, e.g. Borda Count, Copeland Rule, and Plurality Voting strategies; *consensus-based* (or *democratic*) strategies, which average somehow all the available choices, e.g. Additive Utilitarian, Average without Misery, and Fairness strategies; and *borderline* strategies, also called *role-based* strategies in [5], which only consider a subset of choices based on user roles or any other relevant criterion, e.g. Dictatorship, Least Misery and Most Pleasure strategies.

In [17], Mathoff presents and empirically evaluates a number of social choice strategies in a TV item recommendation scenario with a small group of users. Here, we summarise such strategies, and cite representative recommender systems that exploit them. In the following, we assume a user has a preference (utility) for each item represented in the form of a numeric 1-10 rating. In all the cases, the greater the rating value, the most useful the item is for the user.

- **Additive utilitarian strategy.** Preference values from group members are added, and the larger the sum the more influential the item is for the group (Table 7.1). Note that the resulting group ranking will be exactly the same as that obtained taking the average of the individual preference values. A potential problem of this strategy is that individuals' opinions tend to be less significant as larger the group is. This strategy could also use a weighted schema, where a weight is attached to individual preferences depending on multiple criteria for single or multiple users. For example, in *INTRIGUE* [2], *weights* are assigned to particular users' ratings depending on the number of people in the group, and the group's members' relevance (children and disabled have a higher relevance).

Table 7.1 Group choice selection following the additive utilitarian strategy. The ranked list of items for the group would be $(i_5 - i_6, i_8, i_4 - i_{10}, i_1, i_9, i_2, i_7, i_3)$

	Item									
User	i_1	i_2	i_3	i_4	i_5	i_6	i_7	i_8	i_9	i_{10}
u_1	10	4	3	6	10	9	6	8	10	8
u_2	1	9	8	9	7	9	6	9	3	8
u_3	10	5	2	7	9	8	5	6	7	6
group	**21**	**18**	**13**	**22**	**26**	**26**	**17**	**23**	**20**	**22**

- **Multiplicative utilitarian strategy.** Instead of adding the preferences, they are multiplied, and the larger the product the more influential the item is for the group

(Table 7.2). This strategy could be self-defeating: in a small group, the opinion of each individual may have too much impact on the product.

Table 7.2 Group choice selection following the multiplicative utilitarian strategy. The ranked list of items for the group would be $(i_6, i_5, i_8, i_{10}, i_4, i_9, i_2 - i_8, i_1, i_3)$

	Item									
User	i_1	i_2	i_3	i_4	i_5	i_6	i_7	i_8	i_9	i_{10}
u_1	10	4	3	6	10	9	6	8	10	8
u_2	1	9	8	9	7	9	6	9	3	8
u_3	10	5	2	7	9	8	5	6	7	6
group	**100**	**180**	**48**	**378**	**630**	**648**	**180**	**432**	**210**	**284**

- **Average strategy.** In this strategy, the group rating for a particular item is computed as the average rating over all individuals (Table 7.3). Note that if no user or item weighting is conducted, the ranking list of this strategy is the same as that of the Utilitarian strategy. *Travel Decision Forum* [15] implements multiple group modelling strategies, including the average strategy and the median strategy, which uses the middle value of the group members' ratings, instead of the average value. In [36], Yu and colleagues present a TV program recommender that performs a variation of the average strategy, where the group preference vector minimises its distance compared to the individual members' preference vectors.

Table 7.3 Group choice selection following the average strategy. The ranked list of items for the group would be $(i_5 - i_6, i_8, i_4 - i_{10}, i_1, i_9, i_2, i_7, i_3)$

	Item									
User	i_1	i_2	i_3	i_4	i_5	i_6	i_7	i_8	i_9	i_{10}
u_1	10	4	3	6	10	9	6	8	10	8
u_2	1	9	8	9	7	9	6	9	3	8
u_3	10	5	2	7	9	8	5	6	7	6
group	**7**	**6**	**4.3**	**7.3**	**8.7**	**8.7**	**5.7**	**7.7**	**6.7**	**7.3**

- **Average without misery strategy**. As the average strategy, this one assigns an item the average of its ratings in the individual profiles. The difference here is that those items which have a rating under a certain threshold will not be considered in the group recommendations. Table 7.4 shows an example of group formation following this strategy with a threshold value of 3. *MusicFX* [21], which chooses a radio station for background music in a fitness centre, follows an average without misery strategy, and a weighted random selection is made from the top stations in order to avoid starvation and always picking the same station. *CATS* system [22] helps users to choose a joint holiday based on individuals' critiques on holiday package features, and applying the misery aspect.

Table 7.4 Group choice selection following the average without misery strategy. The ranked list of items for the group would be $(i_5 - i_6, i_8, i_4 - i_{10}, i_2, i_7)$

	Item									
User	i_1	i_2	i_3	i_4	i_5	i_6	i_7	i_8	i_9	i_{10}
u_1	10	4	3	6	10	9	6	8	10	8
u_2	1	9	8	9	7	9	6	9	3	8
u_3	10	5	2	7	9	8	5	6	7	6
group	-	**18**	-	**22**	**26**	**26**	**17**	**23**	-	**22**

- **Least misery strategy**. The score of an item in the group profile is the minimum of its ratings in the user profiles. The lower rating the less influential the item is for the group. Thus, a group is as satisfied as its least satisfied member (Table 7.5). *PolyLens* [23] uses this strategy, assuming a group of people going to watch a movie together tends to be small, and the group is as happy as its least happy member. Note that a minority of the group could dictate the opinion of the group: although many members like a certain item, if one member really hates it, the preferences associated to it will not appear in the group profile.

Table 7.5 Group choice selection following the least misery strategy. The ranked list of items for the group would be $(i_6, i_5, i_4 - i_8 - i_{10}, i_7, i_2, i_9, i_3, i_1)$

	Item									
User	i_1	i_2	i_3	i_4	i_5	i_6	i_7	i_8	i_9	i_{10}
u_1	10	4	3	6	10	9	6	8	10	8
u_2	1	9	8	9	7	9	6	9	3	8
u_3	10	5	2	7	9	8	5	6	7	6
group	**1**	**4**	**2**	**6**	**7**	**8**	**5**	**6**	**3**	**6**

- **Most pleasure strategy.** It works as the least misery strategy, but instead of considering for an item the smallest ratings of the users, it selects the greatest ones. The higher rating the more influential the item is for the group, as shown in Table 7.6.

Table 7.6 Group choice selection following the least misery strategy. The ranked list of items for the group would be $(i_1 - i_5 - i_9, i_2 - i_4 - i_6 - i_8, i_3 - i_{10}, i_7)$

	Item									
User	i_1	i_2	i_3	i_4	i_5	i_6	i_7	i_8	i_9	i_{10}
u_1	10	4	3	6	10	9	6	8	10	8
u_2	1	9	8	9	7	9	6	9	3	8
u_3	10	5	2	7	9	8	5	6	7	6
group	**10**	**9**	**8**	**9**	**10**	**9**	**6**	**9**	**10**	**8**

- **Fairness strategy.** In this strategy, the items that were rated highest and cause less misery to all the users of the group are combined as follows. A user is randomly selected. His L top rated items are taking into account. From them, the item that less misery causes to the group (that from the worst alternatives that has the highest rating) is chosen for the group profile with a score equal to N, i.e., the number of items. The process continues in the same way considering the remaining $N-1$, $N-2$, etc. items and uniformly diminishing to 1 the further assigned scores. In the final list, the higher score the more influential the item is for the group. Note that this list would be different if we let other users to choose first. To better understand the strategy, let us explain its first step on the example shown in Table 7.7. Suppose we start with user u1, whose top ranked items are i_1, i_5 and i_9. From these items, we choose item i_5 because it is the one that less misery causes to users u_2 and y_3, whose lowest ratings for items i_1, i_5 and i_9 are respectively 1, 7 and 3. We assign item d_5 a score of 10.

Table 7.7 Group choice selection following the fairness strategy. The ranked list of items for the group could be $(i_5, i_6, i_4, i_8, i_{10}, i_7, i_1, i_2, i_9, i_3)$, following the user selecting order u_1, u_2 and u_3, and setting $L = 3$

	Item									
User	i_1	i_2	i_3	i_4	i_5	i_6	i_7	i_8	i_9	i_{10}
u_1	10	4	3	6	10	9	6	8	10	8
u_2	1	9	8	9	7	9	6	9	3	8
u_3	10	5	2	7	9	8	5	6	7	6
group	**4**	**3**	**1**	**8**	**10**	**9**	**5**	**7**	**2**	**6**

- **Plurality voting strategy.** This method follows the same idea of the fairness strategy, but instead of selecting from the L top preferences the one that least misery causes to the group, it chooses the alternative which most votes have obtained. Table 7.8 shows an example of the group formation obtained with the plurality voting strategy. The item ratings involved in the first step of the algorithm are coloured.

Table 7.8 Group choice selection following the plurality voting strategy. The ranked list of items for the group could be (i_5, i_6, i_4, i_8, i_{10}, i_1, i_9, i_2, i_7, i_3), following the user selecting orderu_1, u_2 and u_3, and setting $L = 3$

	Item									
User	i_1	i_2	i_3	i_4	i_5	i_6	i_7	i_8	i_9	i_{10}
u_1	10	4	3	6	10	9	6	8	10	8
u_2	1	9	8	9	7	9	6	9	3	8
u_3	10	5	2	7	9	8	5	6	7	6
group	**5**	**3**	**1**	**8**	**10**	**9**	**2**	**7**	**4**	**6**

- **Approval voting strategy.** A threshold is considered for the item ratings: only those ratings greater or equal than the threshold value are taking into account for the profile combination. An item receives a vote for each user profile that has its rating surpassing the established threshold. The larger the number of votes the more influential the item is for the group (Table 7.9). This strategy intends to promote the election of moderate alternatives: those that are not strongly disliked.

Table 7.9 Group choice selection following the approval voting strategy. The ranked list of items for the group would be ($i_4 - i_5 - i_6 - i_8 - i_{10}$, $i_1 - i_7 - i_9$, $i_2 - i_3$)

	Item									
User	i_1	i_2	i_3	i_4	i_5	i_6	i_7	i_8	i_9	i_{10}
u_1	10	4	3	6	10	9	1	8	10	8
u_2	1	9	8	9	7	9	6	9	3	8
u_3	10	5	2	7	9	8	5	6	7	6
					⇓ *Threshold* = 5					
	Item									
User	i_1	i_2	i_3	i_4	i_5	i_6	i_7	i_8	i_9	i_{10}
u_1	1			1	1	1	1	1	1	1
u_2		1	1	1	1	1	1	1		1
u_3	1			1	1	1		1	1	1
group	**2**	**1**	**1**	**3**	**3**	**3**	**2**	**3**	**2**	**3**

- **Borda count strategy [6].** Scores are assigned to the items according to their ratings in a user profile: those with the lowest value get zero scores, the next one up one point, and so on. When an individual has multiple preferences with the same rating, the averaged sum of their hypothetical scores are equally distributed to the involved items. With the obtained scores, an additive strategy is followed, and the larger the sum the more influential the item is for the group. Table 7.10 shows an example of the two steps followed by Borda count strategy. In the first step, ratings are normalised according to their relative relevance within the users' preferences. The items with the three lowest ratings for user u_1 are coloured in the tables. For the first one (in increasing rating value), d_3, a zero score is assigned. The second one, d_2, receives a score of value 1. The next score to be assigned would be 2. In this case, the next two items with lowest rating value, d_4 and d_7, have the same rating. In this case, two scores (2 and 3) are considered, and the average of them, i.e., $(2+3)/2 = 2.5$, is assigned to both items.

Table 7.10 Group choice selection following the Borda count strategy. The ranked list of items for the group would be $(i_6, i_5, i_1, i_4 - i_8, i_9, i_{10}, i_2, i_7, i_3)$

	Item									
User	i_1	i_2	i_3	i_4	i_5	i_6	i_7	i_8	i_9	i_{10}
u_1	10	4	3	6	10	9	6	8	10	8
u_2	1	9	8	9	7	9	6	9	3	8
u_3	10	5	2	7	9	8	5	6	7	6

⇓

	Item									
User	i_1	i_2	i_3	i_4	i_5	i_6	i_7	i_8	i_9	i_{10}
u_1	8	1	0	2.5	8	6	2.5	4.5	8	4.5
u_2	0	7.5	4.5	7.5	3	7.5	2	7.5	1	4.5
u_3	9	1.5	0	5.5	8	6	1.5	3.5	5.5	3.5
group	**17**	**10**	**4.5**	**15.5**	**19**	**20.5**	**6**	**15.5**	**14.5**	**12.5**

- **Copeland rule strategy [9].** Being a form of majority voting, this strategy sorts the items according to their *Copeland index*: the difference between the number of times an item beats (has higher ratings) the rest of the items and the number of times it loses. Table 7.11 shows an example of Copeland rule strategy. In the bottom table, a $+/-$ symbol in the ij-th cell (i for rows, and j for columns) means that item at j-th column was rated higher/lower than item at i-th row by the majority of the users. A zero value in a cell means that the corresponding items were rated with the same number of "beats" and "looses".

Table 7.11 Group choice selection following the Copeland rule strategy. The ranked list of items for the group would be $(i_5, i_1, i_6, i_9, i_4, i_8, i_{10}, i_2, i_7, i_3)$

	Item									
User	i_1	i_2	i_3	i_4	i_5	i_6	i_7	i_8	i_9	i_{10}
u_1	10	4	3	6	10	9	6	8	10	8
u_2	1	9	8	9	7	9	6	9	3	8
u_3	10	5	2	7	9	8	5	6	7	6

⇓

	Item									
User	i_1	i_2	i_3	i_4	i_5	i_6	i_7	i_8	i_9	i_{10}
u_1	0	-	-	-	0	-	-	-	0	-
u_2	+	0	-	+	+	+	0	+	+	+
u_3	+	+	0	+	+	+	+	+	+	+
u_4	+	-	-	0	+	+	-	0	0	-
u_5	0	-	-	-	0	-	-	-	-	-
u_6	+	-	-	-	+	0	-	-	-	-
u_7	+	0	-	+	+	+	0	+	+	+
u_8	+	-	-	0	+	+	-	0	+	-
u_9	0	-	-	0	+	+	-	-	0	-
u_{10}	+	-	-	+	+	+	-	+	+	0
group	**+7**	**-6**	**-9**	**+1**	**+8**	**+5**	**-6**	**0**	**+3**	**-3**

7.3 Group Recommender Systems

As stated by several authors [4, 7, 27], group recommender systems can be classified into two main categories: *aggregated models*, which aggregate individual user data into a group data, and generate predictions based on the group data; and *aggregated predictions*, which aggregate the predictions for individual users into group predictions. Other authors [10] have considered the way in which individual preferences are obtained (by content-based or collaborative filtering) as an additional dimension to be taken into account in such categorisation. In any of the above cases, the mechanisms in which user profile models or item predictions are aggregated are manifold, and can be based on any of the social choice strategies explained in Section 2.

In this section, we revise state of the art group recommendation approaches based on user model aggregation, and approaches based on prediction aggregation. We also briefly discuss approaches according to how groups are formed, and approaches that incorporate cooperative consensus to achieve a final recommendation policy agreed by the different members of a group.

7.3.1 Group Recommendation Based on Model Aggregation

The group modelling problem has been addressed by **merging similar individual user profiles**. In this scenario, user profiles are usually represented as sets of weighted preferences or as sets of personal scores assigned by the users to the existing items.

INTRIGUE [2] is a tourist information server that presents information about the area around Torino, Italy. The system recommends sightseeing destinations and itineraries by taking into account the preferences of heterogeneous tourist groups, explains the recommendations by addressing the group members' requirements, and provides an interactive agenda for scheduling a tour. For each individual attraction, a record in a database stores characteristics and properties as a set of feature/value pairs, some of them related to geographical information and others used for matching preferences and interests of the users. Group recommendations are conducted in three steps. Firstly, the group is modelled as a set partitioned into a number of homogeneous subgroups, whose members have similar characteristics and preferences, and are assigned different degrees of influence on the estimation of the group preferences. Next, items are separately ranked by taking the preferences of each subgroup into account. Finally, subgroup-related rankings are merged to obtain the ranking suitable for the whole group.

In [17], Masthoff discusses several strategies based on social choice theory for merging individual user models to adapt to groups (see Section 2). Considering a list of TV programs, a group of viewers represent their interests with sets of personal 1-10 rating for the different TV programs. The author investigates how humans select a sequence of items for the group to watch, how satisfied people believe they would be with the sequence chosen by the different strategies, and how their satisfactions correspond with that predicted by a number of satisfaction functions. These evaluation functions are modified in [18], where satisfaction is modelled as a mood, and assimilation and decline of emotions with time is incorporated. Conducting a user study, she found that participants cared about fairness, and about preventing misery and starvation, as done in strategies like Average, Average without Misery, and Least Misery.

A more sophisticated strategy to merge various individual user profiles based on total distance minimisation is presented in [36]. The authors present a TV program recommender system for multiple viewers, in which the minimisation of the total distance between user profiles guarantees that the merged result could be close to most users' preferences. The shown experimental results prove that the resultant group profile actually reflects most members' preferences of the group.

An evaluation of profile aggregation strategies on a real large-scale dataset of TV viewings is presented in [27], showing that consensus-based strategies (especially the Utilitarian/Average strategy) provided the best recommendation results by comparing the built group profiles to a reference group profile obtained by directly analysing group consumptions.

In [7], we present an approach to automatically identify communities of interest from the tastes and preferences expressed by users in personal ontology-based profiles. The proposed strategy clusters those semantic profile components shared by the users, and according to the found clusters, several layers of interest networks are built. The social relations of these networks are finally used to provide group-oriented recommendations. In this context, we evaluate our approach by using different social choice strategies and, similarly to Senot and colleagues [27], found that consensus-based approaches outperformed borderline strategies, such as Least Misery, Most Pleasure and Plurality Voting strategies.

7.3.2 Group Recommendation Based on Prediction Aggregation

In addition to group modelling, there exist several approaches that have been applied to the problem of making recommendations for groups of people under the framework of **aggregating lists of recommendations** for individual users belonging to the group. For them, we can distinguish two main strategies, namely *collaborative filtering* and *rank aggregation.*

In *collaborative filtering*, a user provides ratings to items, and these ratings are used to suggest her ranked lists that contain other items according to the overall preferences of people with similar rating patterns. Similarity rating patterns are calculated by using different metrics, such as Pearson and Spearman's correlations, and cosine-based distance.

In [14], a video recommender system is presented. Under a client/server architecture, the system receives and sends emails to obtain user ratings, and to provide video suggestions. The recommendations are shown to the users sorted by predicted ratings, and classified by video categories. The system also provides ranked lists from the most similar users, giving thus recommendations to a group of users (virtual community), instead of to individual users. The authors obtained open ended feedback from users indicating interest in establishing direct social contacts within their virtual community.

PolyLens [23] is a collaborative filtering system that suggests movies to groups of people with similar interests, which are expressed through personal five-start scale ratings from the well-known *MovieLens* recommender system [13]. In *PolyLens*, groups of people are explicitly created by users. For each member of a group, a ranked list of movies is obtained from a classic collaborative filtering mechanism. The individual ranked lists are merged according to the least misery principle, i.e., using a social value function where the group's happiness is the minimum of the individual members' happiness scores. Experimenting with *PolyLens*, the authors analysed primary design issues for group recommenders, such as the nature of the groups (in terms of persistency and privacy), the rights of group members, the social value functions for groups, and the interfaces for displaying group recommendations. They found that users not only valued group recommendations, but also were willing to yield some privacy to get the benefits of such recommendations, and extend the recommender system to enable them to invite non-members to participate, via email.

In *rank aggregation*, item recommendation lists are generated for each individual, and afterwards are merged into a single recommendation list for the group. Analogously to model aggregation approaches, different social choice strategies can be used to combine several rankings.

By exploring rank aggregation techniques on *MovieLens* dataset, Baltrunas and colleagues [4] showed that the effectiveness of group does not necessarily decrease when the group size groups, especially if the group have similar minded users. Moreover, they found that if individual recommendations are not correctly ranked (i.e. are not good enough), then recommending items ordered for a group can improve the effectiveness of individual recommendations.

This last result was also presented in [5]. The authors empirically evaluated a number of model aggregation and rank-based prediction aggregation techniques. By using a dataset of explicit ratings for recipes, provided by families of users in an e-health portal, they observed that (i) aggregating individual user models was superior to aggregating individual recommendations, and (ii) role-based weighting outperformed uniform weighting.

7.3.3 Group Formation

Many studies have examined systems that support group formation. The groups can be built **intentionally** (by explicit definition from the users) or **non-intentionally** (by automatic identification from the system).

Kansas [28] is a virtual world in which *a user can explicitly join a group* by moving towards other users, who share a specific virtual spatial region to work collaboratively in a common task. Inside a group, the users can play different roles according to their current capabilities, which are defined by system treatments of user inputs and outputs. These capabilities can be manually acquired and dropped, or can be transferred by one user to another. The authors explain how direct manipulation and control, the "desktop metaphor", might be an interesting approach for human computer interaction in cooperative environments.

MusicFX [21] enables *automatic group formation* by selecting music in a corporate gym according to the musical preferences of people working out at a given time. Thus, performing as a group preference arbitration system, *MusicFX* allows users to influence, but not directly control, the selection of music in the fitness centre. Specifically, each user specifies his preference for each musical genre. An individual preference rating for a genre is presented by a number ranging from -2 to +2. The group preference for that genre is then computed by the sum of the current users' individual preferences. The system uses a weighted random selection policy for selecting one of the group top N music genres. One interesting anecdote the authors found with the system was the fact that people began modifying their workout times to arrive at the gym with other people, often strangers, who shared their music tastes.

7.3.4 Cooperative Consensus

In addition to applying an automatic group modelling algorithm, there exist approaches that make use of **consensus mechanisms** to achieve a final item recommendation policy agreed by the different members of a group. Recently, these approaches have also been called *role-based* [5] and *borderline* [27] strategies.

Travel Decision Forum [15] proposes a manual user interest aggregation method for group modelling by (i) allowing the current member optionally to view (and perhaps copy) the preferences already specified by other members, and (ii) mediating user negotiations offering the users proposals and adaptations of their preferences. This method has several advantages, such as saving of effort, learning from other members, and encouraging assimilation to facilitate the reaching of agreement. In this system, neither user profile merging nor recommendation is used.

Collaborative Advisory Travel System, CATS [22], is a cooperative group travel recommender system which aims to help a group of users arrive at a consensus when they need to plan skiing holidays together; each having their own needs and preferences with respect to what constitutes as an ideal holiday for them. CATS system makes use of visual cues to create emphasis and help users locate relevant information, as well as enhance group awareness of each other's preferences and motivational orientations. Individual user models are defined as set of critiques, i.e., restrictions on vacation features that should be satisfied. The system constructs a reliable group-preference model measuring the quality of each vacation package in terms of its compatibility with the restrictions declared by the members of the group.

7.4 Open Research Problems in Group Recommender Systems

Group recommender systems are still a novel research area. There are many challenging problems for further investigation. Masthoff has recently compiled some of such problems in [19]. Here, we summarise them and include others:

- *Dealing with uncertainty and scarcity in user profiles.* Issues like uncertain, non precise user preferences [10], and cold-start situations may also affect the accuracy of recommendations for certain members of a group.
- *Dealing with social dynamics in a group.* Members of a group may have complex social relationships (e.g., distinct roles, compromises, moods, ages) that affect the individuals' satisfaction for group recommendations. Multi-criteria and constrained recommenders may play a key role in such scenario.
- *Recommending item sequences to a group.* Already suggested items may influence the group members' satisfaction with subsequent recommendations.
- *Explaining recommendations to a group.* Showing how satisfied other members of the group are may improve the user's understanding of received recommendations, and may help to make her accepting suggestions of items she does not like. In this context, however, such transparency has to be balanced with aspects like privacy (e.g. to avoid the embarrassment effect) and scrutability. The reader

is referenced to [34] for a detailed explanation of the roles of explanations in recommender systems.

- *Incorporating negotiation mechanisms.* Encouraging and supporting cooperation is a key aspect in many recommender systems. Facilitating a group of users to easily and friendly negotiate a final decision among a set of item recommendations may increase the individuals' satisfaction.
- *Designing user interfaces.* The user interface of a recommender may affect an individual's satisfaction with group recommendations. For example, in a TV show recommendation scenario, showing the current and the next items to be watched could increase the satisfaction of a user who does like the current suggested item, but is really keen on the subsequent one.
- *Evaluating group recommendations.* Better validation of satisfaction functions should be performed. Among other issues, large-scale evaluations [27], and studies on the affect of group size and composition (e.g., diversity of individuals' preferences within a group) have to be conducted [4, 5]. The reader is referenced to [26] for a detailed discussion of evaluation metrics and methodologies for recommender systems.

7.5 Group Recommender Systems for the Social Web

The Social Web is attracting millions of users, who are no longer mere consumers, but also producers of content. Social systems encourage interaction between users and both online content and other users, thus generating new sources of knowledge for recommender systems. The Social Web presents thus new challenges for recommender systems [12]. In the context of group recommendations, we can highlight the following research directions:

- *Developing new applications.* The huge amount and diversity of user generated content available in the Social Web allow investigating scenarios in which a group of individuals is recommended with "social objects" such as photos, music tracks and video clips stored in online multimedia sharing sites; stories, opinions and reviews published in blogs; and like-minded people registered in online social networks. In such applications, user generated content like ratings, tags, posts, personal bookmarks and social contacts could be exploited by novel group recommendation algorithms [12].
- *Dealing with dynamics and diversity of virtual communities.* In online social networks, people tend to reproduce or extend their relations in the real world to the virtual worlds conformed by the social networks. In [32], the authors show that relationship strength can be accurately inferred from models based on profile similarity and interaction activity on online social networks. Based on these findings, group recommender systems could incorporate content and social interests of group members to perform more accurate item suggestions. For such purpose, it would be necessary to investigate large group characteristics that impact individual decisions, and explore new satisfaction and consensus functions that capture social, interest, and expertise (dis)similarity among the members of

a community [11]. With this respect, because of the evolving composition of online communities, analysing and exploiting the time dimension in the above characteristics may play a key role to obtain more accurate recommendations for community members.

- *Incorporating contextual information.* The anytime-anywhere phenomenon is present in any social system and thus, group recommenders for the Social Web should incorporate contextual information [1]. They would have to automatically detect user presence from inputs provided by mobile, sensor and social data sources [32], and adaptively infer the strength of the social connections within the group, in order to provide accurate recommendations.
- *Finding communities of interest.* In the Social Web, it is very often the case that the membership to a community is unknown or unconscious. In many social applications, a person describes her interests and knowledge in a personal profile to find people with similar ones, but she is not aware of the existence of other (directly or indirectly) related interests and knowledge that may be useful to find those people. Furthermore, depending on the context of application, a user can be interested in different topics and groups of people. In both cases, for individual and group recommender systems, a strategy to automatically identify communities of interest could be very beneficial [7].
- *Integrating user profiles from multiple social systems.* Increasingly, users maintain personal profiles in more and more Web 2.0 systems, such as social networking, personal bookmarking, collaborative tagging, and multimedia sharing sites. Recent studies have shown that inter-linked distributed user preferences expressed in several systems not only tend to overlap, but also enrich individual profiles [31, 35]. A challenging problem in the recommender system field is the issue of integrating such sources of user preference information in order to provide the so called cross-domain recommendations [35]. This clearly opens new research opportunities for group recommenders, which e.g. could suggest to a virtual community sharing interests in a particular domain with items belonging to other domain but liked by some of its members, e.g. recommending specific pieces of classical music to a group with interests in 18th century art.

The authors of this chapter have explored some of the above research paths. We have investigated the use of explicit semantic information as an enhanced modelling ground to combine individual user preferences [30]. We have also researched methods to find implicit communities of interest as a form of latent groups, by mining the kind of user input that is commonly available to a recommender system, along with additional semantic data [7]. We have found an inverse role to the usual one for user communities and groups: besides their natural purpose as user aggregation units, groups provide a basis for user model decomposition. We investigated the use of group models as projecting spaces, to produce sections of user interests –subprofiles– by a projection of complete profiles into the subspace induced by the group model. We found that subprofiles enable more focused and in some situations more precise recommendations than user profiles treated as indivisible units.

7.6 Conclusions

With the advent of the Social Web, people more and more often join virtual communities and social networks, and participate in many different types of collaborative systems, such as wiki-style, product reviewing, and multimedia sharing sites, among others. This together with the progressive spreading of ambient intelligence technologies (e.g., location and mobile-based sensors) in open environments bring in new appealing possibilities and problems for the recommender systems research agenda, which are related to suggesting interesting "social objects" (multimedia items, people, events, plans, etc.) to groups of people having explicit or implicitly bonds among them.

In this chapter, we have revisited existing approaches to group recommendations, and have discussed open research problems in the area, extending such discussion towards a number of potential new research directions related to the context of the Social Web. New complexities and compelling perspectives emerge for recommender systems oriented to groups of users of very different nature and size, as the ones currently growing in the Social Web.

Acknowledgements. This work was supported by the Spanish Ministry of Science and Innovation (TIN2008-06566-C04-02), and the Community of Madrid (S2009TIC-1542).

References

1. Adomavicius, G., Tuzhilin, A.: Context-Aware Recommender Systems. In: Ricci, F., Rokach, L., Shapira, B., Kantor, P.B. (eds.) Recommender Systems Handbook, pp. 217–253 (2011)
2. Ardissono, L., Goy, A., Petrone, G., Segnan, M., Torasso, P.: INTRIGUE: Personalized Recommendation of Tourism Attractions for Desktop and Handset Devices. Applied Artificial Intelligence 17(8-9), 687–714 (2003)
3. Baeza-Yates, R., Ribeiro Neto, B.: Modern Information Retrieval. Addison-Wesley (1999)
4. Baltrunas, L., Makcinskas, T., Ricci, F.: Group Recommendations with Rank Aggregation and Collaborative Filtering. In: Proceedings of the 4th ACM Conference on Recommender Systems (RecSys 2010), pp. 119–126 (2010)
5. Berkovsky, S., Freyne, J.: Group-based Recipe Recommendations: Analysis of Data Aggregation Strategies. In: Proceedings of the 4th ACM Conference on Recommender Systems (RecSys 2010), pp. 111–118 (2010)
6. Borda, J.C.: Mémoire sur les Élections au Scrutin. Histoire de l' Académie Royale des Sciences (1781)
7. Cantador, I., Castells, P.: Extracting Multilayered Communities of Interest from Semantic User Profiles: Application to Group Modeling and Hybrid Recommendations. In: Computers in Human Behavior. Elsevier (in Press, 2011)
8. Chao, D.L., Balthrop, J., Forrest, S.: Adaptive Radio: Achieving Consensus using Negative Preferences. In: Proceedings of the 2005 International ACM Conference on Supporting Group Work (GROUP 2005), pp. 120–123 (2005)
9. Copeland, A.H.: A Reasonable Social Welfare Function. In: Seminar on Applications of Mathematics to the Social Sciences, University of Michigan (1951)

10. De Campos, L.M., Fernández-Luna, J.M., Huete, J.F., Rueda-Morales, M.A.: Managing Uncertainty in Group Recommending Processes. User Modeling and User-Adapted Interaction 19(3), 207–242 (2009)
11. Gartrell, M., Xing, X., Lv, Q., Beach, A., Han, R., Mishra, S., Seada, K.: Enhancing Group Recommendation by Incorporating Social Relationship Interactions. In: Proceedings of the 16th ACM International Conference on Supporting Group Work (GROUP 2010), pp. 97–106 (2010)
12. Geyer, W., Freyne, J., Mobasher, B., Anand, S.S., Dugan, C.: 2nd Workshop on Recommender Systems and the Social Web. In: Proceedings of the 4th ACM Conference on Recommender Systems (RecSys 2010), pp. 379–380 (2010)
13. Herlocker, J., Konstan, J.A., Borchers, A., Riedl, J.: An Algorithmic Framework for Performing Collaborative Filtering. In: Proceedings of the 22nd ACM Conference on Research and Development in Information Retrieval (SIGIR 1999), pp. 230–237 (1999)
14. Hill, W., Stead, L., Rosenstein, M., Furnas, G.: Recommending and Evaluating Choices in a Virtual Community of Use. In: Proceedings of the 13th International Conference on Human Factors in Computing Systems (CHI 1995), pp. 194–201 (1995)
15. Jameson, A.: More than the Sum of its Members: Challenges for Group Recommender Systems. In: Proceedings of the International Working Conference on Advanced Visual Interfaces (AVI 2004), pp. 48–54 (2004)
16. Jameson, A., Smyth, B.: Recommendation to Groups. In: Brusilovsky, P., Kobsa, A., Nejdl, W. (eds.) Adaptive Web 2007. LNCS, vol. 4321, pp. 596–627. Springer, Heidelberg (2007)
17. Masthoff, J.: Group Modeling: Selecting a Sequence of Television Items to Suit a Group of Viewers. User Modeling and User-Adapted Interaction 14(1), 37–85 (2004)
18. Masthoff, J.: The Pursuit of Satisfaction: Affective State in Group Recommender Systems. In: Ardissono, L., Brna, P., Mitrović, A. (eds.) UM 2005. LNCS (LNAI), vol. 3538, pp. 297–306. Springer, Heidelberg (2005)
19. Masthoff, J.: Group Recommender Systems: Combining Individual Models. In: Ricci, F., Rokach, L., Shapira, B., Kantor, P.B. (eds.) Recommender Systems Handbook, pp. 677–702 (2011)
20. McCarthy, J.F.: Pocket RestaurantFinder: A Situated Recommender System for Groups. In: Proceedings of the ACM CHI 2002 International Workshop on Mobile Ad-Hoc Communication (2002)
21. McCarthy, J.F., Anagnost, T.D.: MusicFX: An Arbiter of Group Preferences for Computer Supported Collaborative Workouts. In: Proceedings of the 1998 ACM Conference on Computer Supported Cooperative Work (CSCW 1998), pp. 363–372 (1998)
22. McCarthy, K., Salamo, M., McGinty, L., Smyth, B.: CATS: A Synchronous Approach to Collaborative Group Recommendation. In: Proceedings of the 19th International Florida Artificial Intelligence Research Society Conference (FLAIRS 2006), pp. 1–16 (2006)
23. O'Connor, M., Cosley, D., Konstan, J.A., Riedl, J.: PolyLens: A Recommender System for Groups of Users. In: Proceedings of the 7th European Conference on Computer Supported Cooperative Work (ECSCW 2001), pp. 199–218 (2001)
24. Pattanaik, P.K.: Voting and Collective Choice. Cambridge University Press (2001)
25. Pizzutilo, S., De Carolis, B., Cozzolongo, G., Ambruoso, F.: Group Modeling in a Public Space: Methods, Techniques, Experiences. In: Proceedings of the 5th WSEAS International Conference on Applied Informatics and Communications (AIC 2005), pp. 175–180 (2005)
26. Shani, G., Gunawardana, A.: Evaluating Recommendation Systems. In: Ricci, F., Rokach, L., Shapira, B., Kantor, P.B. (eds.) Recommender Systems Handbook, pp. 257–297 (2011)

27. Senot, C., Kostadinov, D., Bouzid, M., Picault, J., Aghasaryan, A., Bernier, C.: Analysis of Strategies for Building Group Profiles. In: De Bra, P., Kobsa, A., Chin, D. (eds.) UMAP 2010. LNCS, vol. 6075, pp. 40–51. Springer, Heidelberg (2010)
28. Smith, R.B., Hixon, R., Horan, B.: Supporting Flexible Roles in a Shared Space. In: Proceedings of the 1998 ACM Conference on Computer Supported Cooperative Work (CSCW 1998), pp. 197–206 (1998)
29. Smyth, B., Balfe, E., Freyne, J., Briggs, P., Coyle, M., Boydell, O.: Exploiting Query Repetition and Regularity in an Adaptive Community-Based Web Search Engine. User Modeling and User-Adapted Interaction 14(5), 383–423 (2005)
30. Szomszor, M., Alani, H., Cantador, I., O'Hara, K., Shadbolt, N.R.: Semantic Modelling of User Interests Based on Cross-Folksonomy Analysis. In: Sheth, A.P., Staab, S., Dean, M., Paolucci, M., Maynard, D., Finin, T., Thirunarayan, K. (eds.) ISWC 2008. LNCS, vol. 5318, pp. 632–648. Springer, Heidelberg (2008)
31. Szomszor, M., Cantador, I., Alani, H.: Correlating User Profiles from Multiple Folksonomies. In: Proceedings of the 19th ACM Conference on Hypertext and Hypermedia (Hypertext 2008), pp. 33–42 (2008)
32. Szomszor, M., Cattuto, C., Van den Broeck, W., Barrat, A., Alani, H.: Semantics, Sensors, and the Social Web: The Live Social Semantics Experiments. In: Aroyo, L., Antoniou, G., Hyvönen, E., ten Teije, A., Stuckenschmidt, H., Cabral, L., Tudorache, T. (eds.) ESWC 2010. LNCS, vol. 6089, pp. 196–210. Springer, Heidelberg (2010)
33. Taylor, A.D.: Mathematics and Politics: Strategy, Voting, Power and Proof (1995)
34. Tintarev, N., Masthoff, J.: Designing and Evaluating Explanations for Recommender Systems. In: Ricci, F., Rokach, L., Shapira, B., Kantor, P.B. (eds.) Recommender Systems Handbook, pp. 479–510 (2011)
35. Winoto, P., Ya Tang, T.: If You Like the Devil Wears Prada the Book, Will You also Enjoy the Devil Wears Prada the Movie? A Study of Cross-Domain Recommendations. New Generation Computing 26(3), 209–225 (2008)
36. Yu, Z., Zhou, X., Hao, Y., Gu, J.: TV Program Recommendation for Multiple Viewers Based on user Profile Merging. User Modeling and User-Adapted Interaction 16(1), 63–82 (2006)

Chapter 8
Augmenting Collaborative Recommenders by Fusing Social Relationships: Membership and Friendship

Quan Yuan, Li Chen, and Shiwan Zhao

Abstract. Collaborative filtering (CF) based recommender systems often suffer from the sparsity problem, particularly for new and inactive users when they use the system. The emerging trend of social networking sites can potentially help alleviate the sparsity problem with their provided social relationship data, by which users' similar interests might be inferred even with few of their behavioral data with items (e.g., ratings). Previous works mainly focus on the friendship and trust relation in this respect. However, in this paper, we have in-depth explored a new kind of social relationship - the membership and its combinational effect with friendship. The social relationships are fused into the CF recommender via a graph-based framework on sparse and dense datasets as obtained from Last.fm. Our experiments have not only revealed the significant effects of the two relationships, especially the membership, in augmenting recommendation accuracy in the sparse data condition, but also identified the outperforming ability of the graph modeling in terms of realizing the optimal fusion mechanism.

8.1 Introduction

In recent years, collaborative-filtering (CF) based recommender systems have been widely developed in order to effectively support users' decision-making process especially when they are confronted with overwhelming information in the current

Quan Yuan
IBM Research - China, Zhongguancun Software Park, Haidian District, Beijing, China
e-mail: quanyuan@cn.ibm.com

Li Chen
Department of Computer Science, Hong Kong Baptist University, Hong Kong
e-mail: lichen@comp.hkbu.edu.hk

Shiwan Zhao
IBM Research - China, Zhongguancun Software Park, Haidian District, Beijing, China
e-mail: zhaosw@cn.ibm.com

J.J. Pazos Arias et al.: Recommender Systems for the Social Web, ISRL 32, pp. 159–175.
springerlink.com

Web environment. There are two basic entities considered by the recommender: the user and the item. The user provides his rates on items (e.g. movies, music, books, etc.) that he has experienced, based on which the system can connect him with persons who have similar interests and then recommend to him items that are preferred by these like-minded neighbors. In some cases which do not have explicit rating values available, the user's interaction with items can also be considered (e.g., if a user watched a movie, the "rating" is represented as 1; 0 otherwise). Recommendation based on this implicit feedback is also named as log-based CF [26].

However, it inevitably encounters the sparsity problem in the explicit or implicit rating based CF systems. The fact is that usually only active users give ratings to a few proportion of total items (which will become the popular items), while new or inactive users barely give ratings, which phenomenon is particularly obvious when the system is in the initial stage of use. Because of the insufficient information in the dataset, it is hard to improve the accuracy of the similarity measures between users, and hence brings less desirable effects on giving recommendations.

The emerging trend of social networking can potentially alleviate the sparsity problem by inferring users' similar interests through their social relationships. As a matter of fact, many websites now support online user communities, that include media-sharing sites (e.g., *Youtube*, *Last.fm*), e-commerce sites (e.g.,*Amazon.com* and *eBay.com*), and discussion groups like Google Groups. Community facilities are provided so that users can create and access to their community information and communicate with their friends or members. For example, on Last.fm (a worldwide popular social music website), the user can establish friendship with others by "finding people" and/or join an interest group with other users who have similar music tastes (e.g. through "finding groups").

Encouraged by the available data sources, researchers and industrial companies have been engaged in fusing such kind of information into traditional CF methods [20], [15], [8], [22]. Amazon has also tapped into Fracebook through its provided API, to obtain users' social profile and social connections, for enhancing product recommendations. Therefore, for new/inactive users who have little data in Amazon, the site can still provide them with friends' recommendations. However, most of current works emphasize on friendship or trust relations (see Related Works). Since Friendship is an inherently ambiguous relational descriptor ([3], [4]) and trust relations is not so widely available on the internet, there is a need to explore other types of popular social relations to improve the recommendation's accuracy, particularly for new/inactive users.

To the best of our knowledge, few researchers have considered the membership for recommendation, though, the action of joining groups might reflect users' interests as well. In addition, few have touched the interplay between various types of social relationships, such as the combination of friendship and membership. Moreover, existing fusion approaches have been mainly based on the explicit rating-matrix to incorporate the social relationship in order to minimize the prediction error on ratings, while less on implicit user data. From algorithm's perspective, alternative approaches should be explored, such as the graph model because it may more effectively accommodate the inherent transitive characteristics of the relationship.

Thus, in this paper, we have mainly focused on the two kinds of relationships: membership and friendship, to not only in-depth explore their respective impacts when being fused into CF, but also study their combined effects for the fusion. Since graph-based algorithm has its advantage in mining user preferences from a global view instead of pair-wise computation, we adopt the graph-based fusion to incorporate the two types of social relationships. The major findings from the experiment can be summarized as follows:

1. Friendship and membership can both effectively boost the accuracy of recommendation in the sparse user data, and the effect of membership outperforms the one of friendship in both fusion frameworks.
2. The combined effect of membership and friendship only exhibits a slight improvement than fusing membership only. For the dense user data, the fusion impact is not obvious, which means the incorporation of membership and/or friendship did not help improve the recommendation accuracy when there are sufficient user records.
3. The graph-based method has been further divided into two sub-approaches while computing the recommendations: one is retrieving recommendations directly on the graph as resulting from the random walk, and another is based on the graph to obtain the set of the nearest neighbors with the current user. The former method is shown most effective in the sparse data when being fused with membership, whereas the latter neighborhood-based method is the best one in the dense dataset.

In the next section, we will first introduce related works, and then give the details of graph fusion method. Then a comparative experiment will be introduced, which identified the effect of two fusion strategies on sparse and dense user data respectively. Finally, we conclude our work and indicate the future directions.

8.2 Related Work

We first review the related literatures from two perspectives: one is from the data's perspective, which means solving the sparsity problem by fusing social data with user-item matrix to improve standard CF; another is from the algorithm's perspective, which reviews related graph-based recommendation algorithms.

8.2.1 Fusion of Social Relationships

Recently, with the increasing development of social websites and appearance of social data, researchers have begun to pay attention to the social data and explored its usage in recommender systems. [18], to our knowledge, first proposed the combination of social networks and collaborative filtering. [8] proposed to use trust in web-based social networks to create predictive movie recommendations. The trust value was concretely obtained by requiring users to specify how much they trust the people they know. Replacing Pearson-correlation with the trust value for similarity measures in collaborating filtering based recommendation computation, they proved that the predictive accuracy is significantly better. Konstas [19] adopted Random Walk with Restart to model the friendship and social annotation (i.e., tagging) in a music track recommendation system. [10] used social network data for neighborhood generation. In a Munich-based German community, friends are compared to neighbors of collaborative filtering for rating prediction. Their results showed that the social friendship can benefit the traditional recommender system. [22] proposed a factor analysis approach based on probabilistic matrix factorization to solve the data sparsity and poor prediction accuracy problems, by employing both the users' social network information and rating records.

Though some of related works also concentrated on using friendship to improve recommendations. However, it has been shown that online friendship sometimes does not work well due to its inherent ambiguity as a relational descriptor [4]. Compared to online friendship, online community membership potentially contains more information about users' preferences. [24] used membership for recommending online communities to members of the Orkut social network. However, their recommendations were on a per-community basis, rather than on a per-user basis. [23] described an implicit social graph which is formed by users' interactions with contacts, and they proposed a friend suggestion algorithm that uses the implicit social graph to recommend friends.

8.2.2 Graph-Based Recommender Algorithm

The computation of user/item similarity plays a key role in user/item-based collaborative recommenders. Popular measurements of user similarity are Cosine similarity and Pearson's correlation coefficient (see [5] for examples). The limitation is that they only use the local pairwise user information for identifying neighborhood.

Recent years, graph-based methods have been introduced to model relations between users and items from a global perspective, and been used to seamlessly incorporate heterogeneous data sources. [1] is the first work we found that applied graph-theoretic approach for CF. Huang proposed a two-level graph model for products [14], in which the two layers of nodes represent products and customers respectively, and three types of links between nodes are: the product-product, the user-user, and the user-product link. The recommendation is generated based on the association strengths between a customer and products.

Random walks on graph have been extensively discussed [6] and shown a rather good performance in the recommendation area. M. Gori and A. Pucci proposed a random-walk based scoring algorithm, ItemRank [9], which can be used to rank products according to expected user preferences, so as to recommend top ranked items to interested users. Similarly, Baluja [2] made video recommendations for *YouTube* through random walk on the view graph, which is a bipartite graph containing users and videos where links are visiting logs of users on videos. F. Fouss [7] presented a new perspective on characterizing the similarity between elements of a database or, more generally, nodes of a weighted and undirected graph. This similarity called L^+, the pseudoinverse of the Laplacian matrix of the graph. Their experimental results on the MovieLens database showed that the Laplacian-based similarity computation performed well in comparison with other methods. [16] proposed TrustWalker, based on a random walk model that combines the trust relation and collaborative filtering for recommendation. Lately, [27] leverages graph to make temporal recommendation via modeling users' short-term and long-term preferences.

However, the limitation of related work in the "fusion of social relationship" (the first subsection) is that few have considered the potential positive role of membership, and its combined effects with other social relationships. On the other hand, in the related work on "graph-based recommender algorithm", no work on its performance as a fusion platform has been conducted. In this paper, we therefore aim at studying the impact of membership on increasing recommendation accuracy, and investigating how to effectively fuse both of friendship and membership into CF algorithm via graph-based random walk approach in order to improve the performance and solve the sparsity problem. To the best of our knowledge, our work is one of the first attempts to use both friendship and membership to enhance recommender systems.

8.3 Fusing Social Relationships into Recommenders

We first illustrate how to construct the graph which contains all the information, and then describe two types of random walk model for making personalized recommendation based on this graph.

8.3.1 Fusing via Graph

Given that social network is inherently in a graph structure with the transitivity characteristic as a key feature to reflect social relations, we are motivated to use graph-based random walk algorithm for modeling these social data (i.e. membership and friendship). We think that the transitivity of the graph would likely help the computation of the similarity between two users. In general, the graph fusion can be summarized into 4 steps, as shown in Figure 1(a): firstly we construct an social heterogeneous graph; we then fuse social relationships on the graph by adjusting the weights of three types of edges, and then conduct two kinds of random walk on the graph; finally, we generate recommendation through the direct recommendation on the graph or neighborhood-based approach to obtain users' nearest neighbors.

8.3.1.1 Graph Construction for Social Community

The first key issue is to construct a social graph which can include all types of heterogeneous social data, and hence reflect the preference similarity between users.

The relationship data in a social community can be interactive relationship: user watched a movie or listened to a song; or be social relationship like membership or friendship. Let us use Last.fm which contains all of these data as an example. It can be modeled by a graph G in the following way: there are three types of nodes in Last.fm, the user, the artist (the item) and the group, and each element of user set, artist set and group set corresponds to a node in the graph; and the interactive relationship (a user *listens_to* an artist), social relationship (a user is a *member_of* a group, or a user is a *friend_of* another user) are expressed as edges.

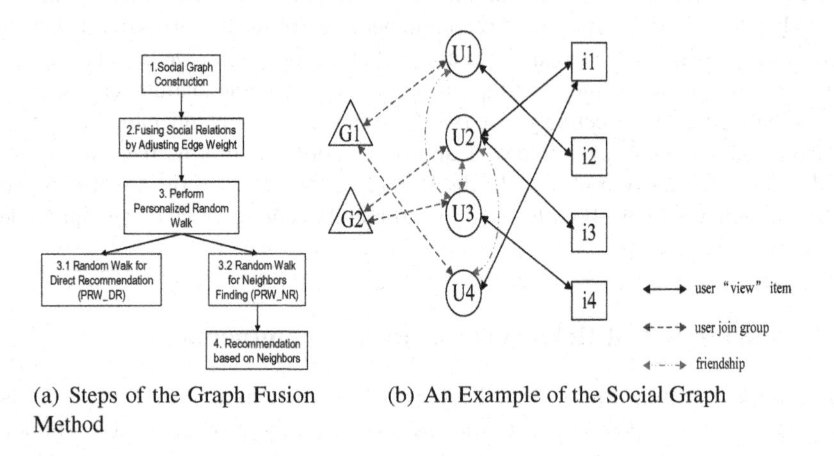

(a) Steps of the Graph Fusion Method

(b) An Example of the Social Graph

Fig. 8.1 Fusing Membership/Friendship via Graph

We represent the structure of a typical social graph as $G(U,I,G,E,w)$, where U denotes the set of user nodes, I is the set of item nodes, and G is the set of group nodes. $w: E \rightarrow \mathbf{R}$ denotes a non-negative weight function for edges. In this example of social graph (see Figure 1(b)), it contains 4 user nodes, 2 group nodes, and 4 item nodes. This example indicates that user u_1 interacted with items i_2, user u_2 interacted with items i_1, i_3, etc. Besides the interaction data, there are two social data: one is the membership, e.g. user u_1 and u_4 joined the same group g_1 and user u_2 and u_3 joined the same group g_2; another is the friendship, e.g. user u_1 and u_3 are friends, u_2 and u_4 are friends, etc.

In order to differentiate the three types of data, we introduce three types of edges accordingly: user-item edge, user-group edge, and user-user edge (referring to the friendship between users), and assign the different edges with different weights, that are w_{ui}, w_{ug} and w_{uu} respectively.

Initially, we set the default weights w^0_{ui}, w^0_{ug} and w^0_{uu} as follows:

For the initial weight of user-item edge (that is w^0_{ui}), we take the number of interactions (e.g. how many times a user listened to a song) into account based on the formula in [13]. The weight w^0_{ui} is concretely defined as:

$$w^0_{ui} = 1 + \rho * \log(1 + n_{ui}/\sigma) \tag{8.1}$$

where n_{ui} is the number of interaction counts, ρ and σ are the parameters. In the simplest situation, $\rho = 0$, $w_{ui} = 1$ for all the user-item edges, which means that the number of interactions is not counted. In our experiments, we set $\rho = 1$ and $\sigma = 1$.

For w^0_{uu}, we adopt the cosine similarity of user-friend matrix as the initial weights of user-user edges. For w^0_{ug}, we set it to 1 if user u joined a group g, otherwise it is 0.

8.3.1.2 Fusion of Social Data by Adjusting Weights of Edges

Firstly, when fusing membership only on the graph, there are only two types of edges on the social graph: user-item edge and user-group edge. Based on their initial weights, we introduce the parameter λ_g to balance the ratios between user-item edge and user-group edge as follows:

$$\begin{aligned} w_{ui} &= w^0_{ui} * (1 - \lambda_g) \\ w_{ug} &= w^0_{ug} * \lambda_g \end{aligned} \tag{8.2}$$

So in this case, the bigger the λ_g is, the bigger the impact of membership brings in making recommendation.

Secondly, when fusing friendship only on the graph, there are user-item edge and user-user edge on the graph. Based on their initial weights, we still use parameter λ_g to balance the ratio between user-item edge and user-user edge as follows:

$$\begin{aligned} w_{ui} &= w^0_{ui} * (1 - \lambda_g) \\ w_{uu} &= w^0_{uu} * \lambda_g \end{aligned} \tag{8.3}$$

Here, the bigger the λ_g is, the bigger the impact of friendship brings in making recommendation.

Finally, when fusing membership and friendship together on the graph, three types of edges are considered simultaneously, and we introduce two parameters λ_g and β_g to balance their ratio as follows:

$$\begin{aligned} w_{uu} &= w^0_{uu} * \lambda_g \\ w_{ug} &= w^0_{ug} * (1 - \lambda_g) * \beta_g \\ w_{ui} &= w^0_{ui} * (1 - \lambda_g) * (1 - \beta_g) \end{aligned} \tag{8.4}$$

In this situation, the bigger the λ_g is, the bigger the impact of friendship takes in making recommendation; and the bigger the β_g is, the bigger the impact of membership is.

After the graph was built, we define the corresponding symmetric adjacency matrix A of graph G, where the element a_{ij} was defined as: $a_{ij} = w_{ij}$ (where w_{ij} is the weight of each edge), if node i is connected to node j otherwise $a_{ij} = 0$.

8.3.1.3 Personalized Random Walk on Social Graph

When the graph construction stage is finished, we run the Personalized Random Walk algorithm (PRW) on the social graph.

Recommendation via the PRW on graphs has been well studied in [9, 21, 17]. The following formulation is used by PageRank [12] to rank nodes in a graph:

$$\mathrm{PR} = \alpha \cdot M \cdot \mathrm{PR} + (1 - \alpha) \cdot d \tag{8.5}$$

where α is the damping factor, M is a transition matrix and vector d is a user-specific personalized vector that indicates which node the random walker should jump to after a restart:

$$d(v) = \begin{cases} 1 & v = v_u \\ 0 & otherwise \end{cases} \tag{8.6}$$

So except on the corresponding position of the current user u for which d(v) is 1, it is 0 on the other positions.

We then use two different strategies for making recommendations based on the random walk.

PRW based Direct Recommendation (PRW-DR): After running PRW, we can get a personalized ranked list of nodes (including user node, item node, and group node) for the current user. Each node has a ranking value, which can be considered as the probability that the current user would likely visit after the random walk. So one strategy is to directly recommend the top-N item nodes (sorted by their rankings) to the current user. We call it DR for short.

So when applying DR on the above-mentioned graph, there are four cases in total: running DR on the graph which is built only from rating matrix (DR); on the graph which is built from rating matrix and membership (DR.M); on the graph that consists of rating matrix and friendship (DR.F); and finally on the graph which includes all the three types of data sources (DR.M+F).

PRW based Neighborhood Recommendation (PRW-NR): Another strategy is to only consider the top-N user nodes according to their rankings as the nearest neighbors of the current user, and then make use of the neighborhood similarities to generate recommendations by aggregating neighbors' viewed items. We name this approach NR. When applying NR on the graph, similar to DR, for the four cases with different types of data sources involved, we use NR, NR.M, NR.F, NR.M+F for short correspondingly.

To the best of our knowledge, this is the first work aimed at uncovering the difference of the two random walk-based recommendation strategies. We will illustrate their performance on dense and sparse datasets respectively later in experiments. Another difference between our approach and previous works on random walk lies in that PRW-DR and PRW-NR both run on a heterogeneous social graph, while the others are mostly on the unipartite graph with only one type of node - user node involved, in which the social data related to more than one type of node (e.g.,group node) can not be easily incorporated in their graphs.

8.3.2 Complexity Analysis

For the graph approach based on Personalized Random Walk, when considering user-item matrix only, since it walks on the global graph, and is implemented in an iterative method, for all users its complexity is $O(m \cdot (e_{ui}) \cdot |U|)$, where m is the iteration times, and e_{ui} is the number of user-item edges. When fusing friendship on the graph, since it only added user-user edges, the total complexity is $O(m \cdot (e_{ui} + e_f) \cdot |U|)$, where e_f is the number of user-user edges. While fusing membership, it added user-group edges on the graph, so the total complexity is $O(m \cdot (e_{ui} + e_{ug}) \cdot |U|)$, where e_{ug} is the number of user-group edges.

8.4 Experiments

8.4.1 Data Sets

Traditional data sets used in the evaluation of collaborative filtering systems, such as Netflix and MovieLens, do not include explicit social relationships, which is why we used *Last.fm* (a worldwide popular social music site) where community information is available so that an entity-relation model can be generated which includes the relationship between users.

For our purpose, we extracted two typical social relationships by accessing the site's Web Service APIs: the membership which describes the user's participations in groups and the friendship between users. Besides, we think it is more meaningful to recommend artists instead of individual songs since user preference on artists would be more constant, so we use artist as the "item" in our recommendations. A user and an item is linked if the user listened to song(s) of the artist, and a user and a community is linked if the user joined the group.

We crawled the user-item interaction data and social relationships from Last.fm, and randomly selected 10128 users and 3859 items from the data. In order to study above algorithms' performance on the data with varied sparsity levels, we sampled a dense set and a sparse set from the original data set. The dense set was made after doing 5-fold splitting on the original user-item data, so it contains 80% of user-item pairs in the training set, and 20% in the test set. The sparsity level of the dense set is 98.54%. The sparse set was made after doing 5-fold splitting on one test set of the dense data, which means it contains 4% (20% / 5) of user-item pairs in the training set, and we use the remaining 96% as the test set. The sparsity level of the sparse set is 99.91%.

Table 8.1 Description of LastFm datasets

element	LastFm
#user	10128
#item	3859
#group	14796
#user-item pair	709740
#user-group pair	97494
#user-user friendship	46617

8.4.2 Evaluation Metrics

We adopted standard metrics in the area of information retrieval to evaluate our recommenders. During each round of cross-validation, we recommended top-N (N=100 in our experiments) artists for each user. We then compared the predicted recommendation list with users' true preferences on artists in the test set, which led to, recall and precision.

1. **Recall.** The score measures the average (on all users) of the proportion (in percentages) of artists from the test sets that appear among the top n ranked list, for some given n. It should be as high as possible for good performance.
2. **Precision.** This metric measures the proportion of recommended items that are ground-truth items.

However, precision based metrics are not very appropriate, as they require knowing which artists are undesired to a user. Listening to an artist is an indication of liking her/him, making recall-oriented measures more applicable [13]. So we mainly consider recall as the primary metric. While we also report the results of precision in the tables.

8.4.3 Results

Since our primary task is to explore the social relationships' ability in solving sparsity problem, we first look into two fusion strategies' performance on the sparse data, and then give their results on the dense data.

8.4.3.1 Results on Sparse Dataset

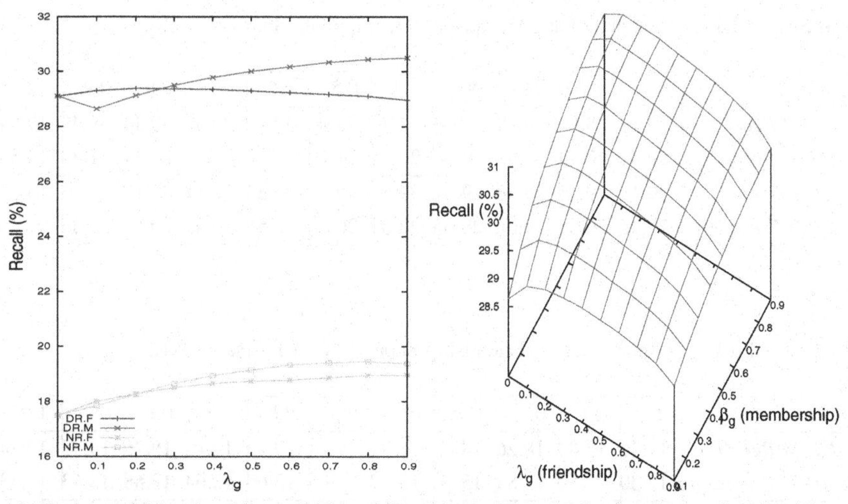

(a) Fusion of Membership and Friendship Separately (b) Fusion of Membership and Friendship Together

Fig. 8.2 Graph Fusion of Membership and Friendship, Sparse data

Figure 2(a) illustrates the fusion results on the graph model in the sparse set condition. First of all, we noticed that in the circumstance of no social relationships, both PRW-based direct recommendation (DR) and neighborhood-based recommendation (NR) boost the baseline from 17.31% (user-based CF by leveraging cosine as similarity measure) to 29.10%, which indicates the impact of graph-based algorithms on mining the inherent linkage based on user-item matrix and hence solving the sparsity problem apparently.

Let us firstly look at direct recommendation (DR)'s performance. When fusing the membership, we get the optimal recall when $\lambda_g = 0.9$ in equation (8.2), and the recall is 30.49%, which increases DR without any social data by 4.78%. As for fusing the friendship, we get the optimal recall (29.38%) when $\lambda_g = 0.2$ in equation (8.3), which is a slight improvement compare to membership's effect.

Then for the Neighborhood-based Recommendation (NR), its performances on fusing both social relations are surprisingly worse than the DR, and almost on the same level as user-based CF. This observation infers that the neighborhood-based approach might be not suitable for the sparse data. The reason is likely because in general, on sparse data, all the neighbors only have very few viewed items in the user-item matrix, so it is hard to get an accurate recommendation that is purely based on the aggregation of neighbors' behaviors.

The following tables (8.2, 8.3) list the detailed results of fusing both relations on the sparse data through two different methods.

Table 8.2 Fusing Membership and Friendship Separately on Sparse via DR

Sparse	0	0.1	0.2	0.3	0.4	0.5	0.6	0.7	0.8	0.9	1.0
DR.M.Recall	29.10	28.64	29.12	29.50	29.78	30.00	30.17	30.33	30.44	**30.49**	19.24
DR.M.Precision	20.46	20.14	20.48	20.74	20.94	21.10	21.22	21.33	21.41	**21.44**	13.53
DR.F.Recall	29.10	29.30	**29.38**	29.37	29.35	29.29	29.23	29.17	29.10	28.96	13.85
DR.F.Precision	20.46	20.61	**20.66**	20.65	20.64	20.60	20.55	20.52	20.47	20.36	9.74

Table 8.3 Fusing Membership and Friendship Separately on Sparse via NR

Sparse	0	0.1	0.2	0.3	0.4	0.5	0.6	0.7	0.8	0.9	1.0
NR.M.Recall	17.49	17.84	18.26	18.65	18.94	19.14	19.33	19.40	**19.45**	19.38	19.24
NR.M.Precision	12.30	12.54	12.84	13.12	13.32	13.46	13.60	13.64	**13.68**	13.63	13.53
NR.F.Recall	17.49	18.00	18.27	18.50	18.65	18.74	18.78	18.86	18.95	**18.96**	13.85
NR.F.Precision	12.30	12.66	12.85	13.01	13.12	13.18	13.20	13.26	13.32	**13.33**	9.74

Finally, we are interested in knowing what the results are when fusing membership and friendship simultaneously on the sparse data. According to equation (8.4), we adopt λ_g and β_g to adjust the weight of three inputs in the graph fusion.

When fusing membership and friendship together on the graph, since DR is much better than NR as shown above, we only draw Figure 2(b) to show the results of DR. We obtain the optimal recall 30.68% when $\lambda_g = 0.1$ and $\beta_g = 0.9$ in equation (8.4). This result outperforms the best result of DR.M (30.49%) a little.

All of the results hence indicate that the best results for the sparse data is achieved when fusing membership and friendship together on the graph and making direct recommendation (DR.M+F), and the runner-up is fusing membership only (DR.M) and the follows friendship only (DR.F). From these results we can see the membership's higher effectiveness in boosting recommendation's accuracy, as well as the advantages of graph-based PRW-DR in solving sparsity problem.

8.4.3.2 Results on Dense Dataset

Besides the sparse data, we also measured the fusion effect of membership and friendship on dense data.

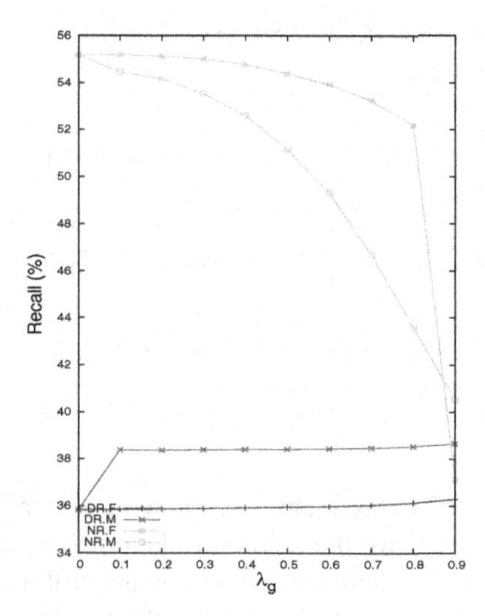

Fig. 8.3 Graph Fusion of Membership and Friendship Separately, Dense data

When we look at Figure 8.3, it is interesting to see that the results of DR and NR have been reversed on this dense data in comparison with their results on sparse data. Although DR still improves the recall after fusing social data, its absolute value is much lower compared with NR, and even worse than the user-based CF without social fusion on this dataset, which is 52.42%.

As for NR, when without fusing any social data, the result is 55.17%, this result improves user-based CF by 5.25%. After fusing the friendship a little, it just slightly increases to 55.19% ($\lambda_g = 0.1$), which is the best result on dense data. While as the friendship increases, the recall begin to decrease. When fusing the membership, the recall begin to reduce from the beginning to the end.

In summary, we see that the best result for the dense data is achieved by NR.F when it is fused with only a small degree of friendship, followed by NR without any social fusion and NR.M, and the last one is DR, which has much lower recall than the above approaches. So we draw the conclusion that social data's is not applicable for dense data in that they already have enough behavioral information. Meanwhile, from the algorithm's perspective, we prove the effectiveness of neighborhood based strategy (NR) in handling dense data, especially when it combines with random walk by leveraging random walk's ability in calculating the neighborhood similarity, and achieving the best result on dense data.

The following tables (8.4, 8.5) list the detail results of fusing both relations on dense data.

Table 8.4 Fusing Membership and Friendship on Dense data via DR

Dense	0	0.1	0.2	0.3	0.4	0.5	0.6	0.7	0.8	0.9	1.0
DR.M.Recall	35.83	38.37	38.36	38.38	38.39	38.40	38.42	38.46	38.52	38.64	**39.85**
DR.M.Precision	5.06	5.42	5.42	5.42	5.42	5.42	5.43	5.43	5.44	5.46	**5.63**
DR.F.Recall	35.83	35.85	35.87	35.89	35.92	35.95	35.98	36.04	36.13	**36.29**	14.06
DR.F.Precision	5.06	5.06	5.07	5.07	5.07	5.08	5.08	5.09	5.10	**5.12**	1.99

Table 8.5 Fusing Membership and Friendship on Dense data via NR

Dense	0	0.1	0.2	0.3	0.4	0.5	0.6	0.7	0.8	0.9	1.0
NR.M.Recall	**55.17**	54.46	54.16	53.53	52.59	51.14	49.31	46.68	43.59	40.53	39.85
NR.M.Precision	**7.79**	7.69	7.65	7.56	7.43	7.22	6.96	6.59	6.16	5.72	5.63
NR.F.Recall	55.17	**55.19**	55.12	55.01	54.77	54.38	53.92	53.24	52.18	37.14	14.06
NR.F.Precision	**7.79**	**7.79**	7.78	7.77	7.73	7.68	7.61	7.52	7.37	5.24	1.99

For the dense data, after experiments, we know that fusing each social relationship separately can not improve the recommendation accuracy, so it is obvious to think that fusing membership and friendship together still has negative effect. Actually, we did this experiment, and it proves the above suppose. Due to the limit of pagesize, we ignore the results here.

8.4.3.3 Overall Accuracy Comparison

In this section, we give the overall accuracy comparison among different fusion algorithms on sparse and dense data, as shown in table 8.6.

On sparse data, the DR.M+F is ranked first, and DR.M is the second one, which proves the effectiveness of membership and graph-based approach on sparse dataset. Moreover, the combined effect of membership and friendship can only slightly improve the result of fusing membership only. Although NR is graph-based approach like DR, NR's poor performance on sparse data tells us that the neighborhood-based approach is not suitable for sparse data since the lack of neighbors' behavioral information can be harmful to the subsequent recommendation process which is based on the aggregation of neighbors' behaviors.

On dense data, the most notable phenomenon is that both membership and friendship are not so effective in augmenting recommendation's accuracy, not to mention their combination. Generally speaking, NR-based approaches are better than DR-based ones, since the first two are both related to neighborhood similarity measure, it implies the effectiveness of neighborhood-based approach in handling dense data given that which contains rich behavioral information for the aggregation and recommendation.

Table 8.6 Optimal recall for each approach

Recall(%)	User-CF	DR	DR.M	DR.F	DR.M+F	NR	NR.M	NR.F
Sparse	17.31	29.1	30.49	29.38	**30.69**	17.49	19.45	18.96
Dense	52.42	35.83	39.85	36.29	-	55.17	54.46	**55.19**

8.5 Conclusion and Future Work

Thus, in this paper, we have targeted to address the sparsity problem for new and inactive users by introducing social relationships, especially the membership that has been few investigated in related literatures. To be concrete, we proposed a graph-based modeling for the fusion of social relationships into collaborative filtering recommenders. In an experiment that has assessed the performance of different fusion methods in both of sparse and dense user data conditions, we have demonstrated the particular effectiveness of social relationships in augmenting recommendations on sparse data, and more notably the distinct role of membership in obtaining the objective. Specifically, the graph-based fusion with membership, which was applied to make direct recommendations, is proven most effective for sparse data. On the other hand, for dense data, the social relationships did not show significant improvement on recommendation accuracy, which may be primarily due to the rating matrix's sufficiency of behavioral info (i.e., users' interaction behaviors with the items). The findings hence verify again the particular impacts of social relationships on addressing the sparsity limitations for new/inactive users.

Driven by the results from this paper regarding the outstanding performance of graph-based modeling in fusing social relations, in the future, we plan to compare the graph method with more alternatives, such as matrix factorization, to further identify its inherent merits. On the other hand, through more experiments, we will consolidate the fusion algorithm's scalability and efficiency in handling with more larger scale social graphs and various product domains.

References

1. Aggarwal, C.C., Wolf, J.L., Wu, K.-L., Yu, P.S.: Horting hatches an egg: a new graph-theoretic approach to collaborative filtering. In: KDD 1999: Proceedings of the Fifth ACM SIGKDD International Conference on Knowledge Discovery and Data Mining, pp. 201-212. ACM, New York (1999)
2. Baluja, S., Seth, R., Sivakumar, D., Jing, Y., Yagnik, J., Kumar, S., Ravichandran, D., Aly, M.: Video suggestion and discovery for youtube: taking random walks through the view graph. In: WWW 2008: Proceeding of the 17th international conference on World Wide Web, pp. 895–904. ACM Press, New York (2008)
3. Baym, N.K., Ledbetter, A.: Tunes that bind?: Predicting friendship strength in a music-based social network. In: AOIR 2008: the Association of Internet Researchers, Copenhagen, Denmark (2008)

4. Boyd, D.: Friends, friendsters, and myspace top 8: Writing community into being on social network sites (2006)
5. Breese, J.S., Heckerman, D., Kadie, C.: Empirical analysis of predictive algorithms for collaborative filtering. In: Proceedings of the 14th Annual Conference on Uncertainty in Artificial Intelligence (UAI 1998), pp. 43–52 (1998)
6. Doyle, P.G., Laurie, S.J.: Random walks and electrical networks (1984)
7. Fouss, F., Pirotte, A., Renders, J.-M., Saerens, M.: Random-walk computation of similarities between nodes of a graph with application to collaborative recommendation, pp. 355-369 (2007)
8. Golbeck, J.: Generating predictive movie recommendations from trust in social networks. Trust Management, 93–104 (2006)
9. Gori, M., Pucci, A.: Itemrank: a random-walk based scoring algorithm for recommender engines. In: IJCAI 2007: Proceedings of the 20th International Joint Conference on Artifical Intelligence, pp. 2766–2771. Morgan Kaufmann Publishers Inc., San Francisco (2007)
10. Groh, G., Ehmig, C.: Recommendations in taste related domains: collaborative filtering vs. social filtering. In: GROUP 2007: Proceedings of the 2007 International ACM Conference on Supporting Group Work, pp. 127–136. ACM Press, New York (2007)
11. Guy, I., Ronen, I., Wilcox, E.: Do you know?: recommending people to invite into your social network. In: IUI 2009: Proceedings of the 13th International Conference on Intelligent User Interfaces, pp. 77–86. ACM Press, New York (2009)
12. Haveliwala, T.H.: Topic-sensitive pagerank. In: WWW 2002: Proceedings of the 11th International Conference on World Wide Web, pp. 517–526. ACM Press, New York (2002)
13. Hu, Y., Koren, Y., Volinsky, C.: Collaborative filtering for implicit feedback datasets. In: ICDM 2008: Proceedings of the 2008 Eighth IEEE International Conference on Data Mining, pp. 263–272. IEEE Computer Society, Washington, DC, USA (2008)
14. Huang, Z.: Graph-based analysis for e-commerce recommendation, Tucson, AZ, USA (2005)
15. Hummel, H.G.K., Berg, B.V.D., Berlanga, A.J., Drachsler, H., Janssen, J., Nadolski, R., Koper, R.: Combining social based and information based approaches for personalised recommendation on sequencing learning activities, pp. 152–168 (2007)
16. Jamali, M., Ester, M.: Trustwalker: a random walk model for combining trust-based and item-based recommendation. In: KDD 2009: Proceedings of the 15th ACM SIGKDD International Conference on Knowledge Discovery and Data Mining, pp. 397–406. ACM Press, New York (2009)
17. Jamali, M., Ester, M.: Trustwalker: a random walk model for combining trust-based and item-based recommendation. In: KDD 2009: Proceedings of the 15th ACM SIGKDD International Conference on Knowledge Discovery and Data Mining, pp. 397–406. ACM Press, New York (2009)
18. Kautz, H., Selman, B., Shah, M.: Referral web: combining social networks and collaborative filtering. ACM Commun., 63–65 (1997)
19. Konstas, I., Stathopoulos, V., Jose, J.M.: On social networks and collaborative recommendation. In: SIGIR 2009: Proceedings of the 32nd International ACM SIGIR Conference on Research and Development in Information Retrieval, pp. 195–202. ACM, New York (2009)
20. Lam, C.: Snack: incorporating social network information in automated collaborative filtering. In: EC 2004: Proceedings of the 5th ACM Conference on Electronic Commerce, pp. 254–255. ACM, New York (2004)

21. Liu, N.N., Yang, Q.: Eigenrank: a ranking-oriented approach to collaborative filtering. In: SIGIR 2008: Proceedings of the 31st Annual International ACM SIGIR Conference on Research and Development in Information Retrieval, pp. 83–90. ACM, New York (2008)
22. Ma, H., Yang, H., Lyu, M.R., King, I.: Sorec: social recommendation using probabilistic matrix factorization. In: CIKM 2008: Proceeding of the 17th ACM Conference on Information and Knowledge Management, pp. 931–940. ACM, New York (2008)
23. Roth, M., Ben-David, A., Deutscher, D., Flysher, G., Horn, I., Leichtberg, A., Leiser, N., Matias, Y., Merom, R.: Suggesting friends using the implicit social graph. In: KDD 2010: Proceedings of the 16th ACM SIGKDD International Conference on Knowledge Discovery and Data Mining, pp. 233–242. ACM, New York (2010)
24. Spertus, E., Sahami, M., Buyukkokten, O.: Evaluating similarity measures: a large-scale study in the orkut social network. In: KDD 2005: Proceedings of the Eleventh ACM SIGKDD International Conference on Knowledge Discovery in Data Mining, pp. 678–684. ACM, New York (2005)
25. Tong, H., Faloutsos, C., Pan, J.-Y.: Fast random walk with restart and its applications. In: ICDM 2006: Proceedings of the Sixth International Conference on Data Mining, pp. 613-622. IEEE Computer Society, Washington, DC, USA (2006)
26. Wang, J., de Vries, A.P., Reinders: A user-item relevance model for log-based collaborative filtering, pp. 37-48 (2006)
27. Xiang, L., Yuan, Q., Zhao, S., Chen, L., Zhang, X., Yang, Q., Sun, J.: Temporal recommendation on graphs via long- and short-term preference fusion. In: KDD 2010: Proceedings of the 16th ACM SIGKDD International Conference on Knowledge Discovery and Data Mining, pp. 723–732. ACM, New York (2010)

Part V
Applications

Chapter 9
Recommendations on the Move

Alicia Rodríguez-Carrión, Celeste Campo, and Carlos García-Rubio

Abstract. Recommender systems can take advantage of the user's current location in order to improve the recommendations about places the user may be interested in. Taking a step further, these suggestions could be based not only on the user's current location, but also on the places where the user is supposed to be in the near future, so the recommended locations would be on the path the user is going to follow. In order to do that we need some location prediction algorithms so that we can get those future locations. In this chapter we explain how to use the algorithms belonging to LZ family (LZ, LeZi Update and Active LeZi) as recommender engines, and we propose some ways of using these algorithms in places where the user has not been before or how to take advantage of the social knowledge about certain place so as to make these recommendations richer. Finally we show a prototype implementation of a recommender system for touristic places made up of these LZ predictors.

9.1 Introduction

A recommender is an application that ranks a set of available choices with respect to certain criteria [11]. An example of recommender system is one that suggests places (shops, restaurants, pubs, museums, historical buildings, etc.) the user may be interested in.

Now that we are immerse in the anywhere, anytime phenomenon, and Internet is more present in mobile devices than ever, recommender systems can be improved by taking into account the current location of the user, so the suggested places are close to the user in the moment she asks for them. Mobile devices offer several ways of locating the user, as we will see in Section 9.2, so this recommendation based on

Alicia Rodríguez-Carrión · Celeste Campo · Carlos García-Rubio
Department of Telematic Engineering
University Carlos III of Madrid
28911 Leganés, Madrid, Spain
e-mail: {arcarrio,celeste,cgr}@it.uc3m.es

J.J. Pazos Arias et al.: Recommender Systems for the Social Web, ISRL 32, pp. 179–193.
springerlink.com

the current location is already possible. This kind of systems is usually known as location-aware recommender systems [10].

Could we make those recommendations even richer? Imagine you need to go shopping for some item. You go to the first shop but, unfortunately, you do not find what you are looking for. At that moment, you probably do not want to know just the shops around your current location, but also the ones close to the places you will visit in the next hours. In other words, it would be very useful if the system could predict where you will go next by analyzing where you have been before, and use this prediction to improve the results offered by the recommender, suggesting you not just the shops of your interest that are close to your current location, but also those close to your next locations and to the routes you will follow.

This chapter is centered in location-aware recommender systems implemented in mobile phones that use location prediction to refine the suggestions provided to the user. In Section 9.2 we present a family of prediction algorithms, called LZ family, used in mobile phones for this purpose. To assess their accuracy, we show location prediction performance measurements obtained using these algorithms with movement traces from real users.

These predictors estimate a user's future location based on past movement patterns of that user. For some applications, for example tourism recommender systems such as the one presented in [8], which are used in places the user has never been before, the recommender cannot make use of the past history of the user to improve the recommendations. Section 9.3 takes care of how to deal with this situation, and also on how to include people's choices in order to add a social component to the recommender system.

Finally, we present a prototype implementation of a recommender system in a mobile phone with location prediction (Section 9.4), and then we sum up the main conclusions of this work together with some issues open to further discussion in Section 9.5.

9.2 Predicting Future Locations in Mobile Phones

In order to make predictions, as well as when making recommendations, there are two key elements without which it would be impossible to predict or recommend: the history about users, which stores past events (actions, choices, locations... depending on the kind of prediction we want to make); and the algorithms which process that history, detect patterns or features modeling the user's behavior and estimate with this information the most probable next action, choice, location... Along this section we are covering some possible data sources that can be used with the prediction algorithms also explained later on.

9.2.1 Location Data Sources

Although Global Positioning System (GPS) is probably the first technology we think about when we want to track users' locations, there are many other ways of doing it. Think about one of those mobile devices that are filling the market nowadays. Besides GPS, they have several data connections as GSM, UMTS, WiFi, Bluetooth, NFC, and also a wide variety of sensors: digital compass, proximity sensors, accelerometers... Therefore we can, for instance, monitor the GSM or UMTS base station (BS) or the WiFi access point (AP) the mobile phone is attached to as the user moves. We can also record the direction the user is heading to using the compass, or the sequence of NFC tags the phone is reading so we can track a user in small indoor environments.

Nevertheless we have only mentioned location data which is directly accessed by the terminal hardware. What about web applications like Foursquare or Facebook Places, with which we can check in the places we stop by? This is a very interesting way of obtaining a record of the sequence of touristic places, shops, restaurants and many other places a user visits during certain period.

The kind of prediction algorithms we are going to consider in next sections is able to deal with some of the data sources described above. This is possible because the predictors work with symbols representing the locations. For example, when using GSM BSs (or cells) we have to translate the cell identifiers into symbols, as shown in Figure 9.1, and then these symbols are the ones feeding the predictors. BSs identifiers can be changed by WiFi APs or NFC tags and the translation is still analogous, as well as for the sequence of places the user checks in using web applications. There are some bibliography where WiFi APs-based location data is used to locate the user and predict her movements [7], and the same applies for cell-based location data case [6].

Therefore a user who takes the path plotted in Figure 9.1 will have a movement or location history, also known as mobility trace, as follows: $L = acdbef$. This trace is one of the main ingredients of a location prediction, since it is the input data of the predictors explained next.

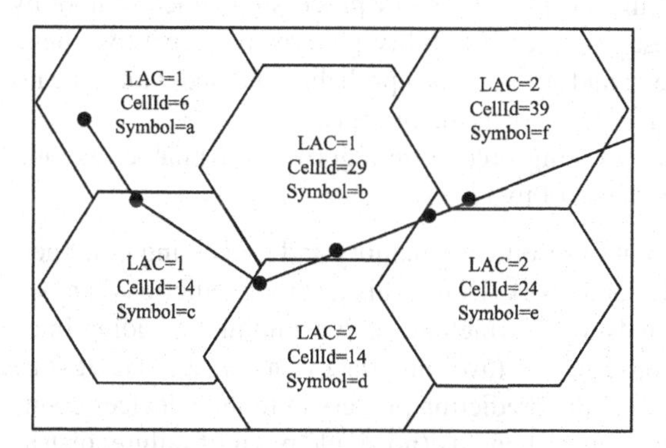

Fig. 9.1 Division made by GSM network, where each cell is identified by its location area code (LAC) and cell identifier (CellId), and its translation into symbols. The solid line represents a user's path and the points show where the user's mobile phone detects every cell change, thus being the places where the trace is updated.

9.2.2 Location Predictors

Location prediction algorithms allow us to detect and learn user's mobility patterns and use this knowledge so as to estimate the most probable next location. Some of these algorithms first analyze several traces, extracting the main features of the users' mobility model and, using these features and the current context, they finally make the predictions. This implies an static behavior, i.e. the algorithm learns the set of features by using a set of past data, and the learned features remains the same during all the prediction process.

On the other hand, there is another kind of algorithms which learn the user mobility model with each new visited location and adapt their predictions to the user's behavior as it changes. This allows more dynamical predictions, meaning that if mobility patterns of a user changes, the algorithm is able to detect those changes and react consequently.

There is an additional classification that can be made attending to the kind of information used by the predictors as input data, resulting into two groups: domain dependent and domain independent algorithms. The first group make their predictions based on location context information such as coordinates, the place function (restaurant, office...) or many others. Domain independent predictors however treat each location as a different symbol, ignoring any other information about that location.

The location predictors belonging to the LZ family which are considered along this chapter belongs to the set of algorithms learning dynamically the user's mobility patterns, and thus adapting to its changes, and also to the group of domain independent algorithms. Because of this last reason the location data sources previously mentioned are very appropriate to provide the sequence of symbols these algorithms need as input data.

When talking about location data, we said that the string of symbols representing the locations the user has passed by is known as movement history or trace, L. The prediction algorithms explained below process this trace symbol by symbol, based on two hypothesis: (i) users' mobility patterns are repetitive, hence the trace is a stationary process; and (ii) there is a probabilistic model underneath the movement, and therefore L is also a stochastic process.

Among all the existing prediction algorithms available, we are focused on LZ family mainly because of two reasons:

- The low amount of resources (memory and processing time) needed by these algorithms in order to work. Due to this fact the predictions can be estimated on the mobile terminals itself, which leads to economic (avoiding Internet connection) and power consumption (avoiding data transferring) savings. Besides these advantages, making the prediction process in mobile devices counts also on all the distributed computing benefits (no single point of failure, distributed computing resources...) as well as on the lack of risk associated with sending sensitive data, as user's location information is, through the network.
- The second advantage of these predictors is derived from belonging to the set of algorithms learning movement patterns dynamically, thus taking into account changes in user's behavior. The mobility model the predictors learn as time goes

by is updated with each new location (i.e. symbol). Therefore if a user at some point changes her routine (stops visiting certain locations, starts visiting new ones, changes the location sequence order...), the algorithm will realize these changes and start making predictions according to the new routine.

Next we are going into the details of each of the three algorithms belonging to the LZ family: LZ, LeZi Update and Active LeZi.

9.2.2.1 LZ Algorithm

This first predictor is the basis for the remaining ones. LZ algorithm [12] takes the input trace L and, considering γ as the empty string, splits L into substrings $s_0 s_1 \dots s_n$ such that $s_0 = \gamma$ and for every $j \geq 1$ the prefix of the substring s_j (i.e. every character of s_j except for the last one) is equal to some previous s_i, $\forall i < j$. As we can see, the division is made in a sequential way, thus considering only the remaining trace after each s_i is determined. For instance, taking a mobility trace $L = ababab cdcbdab$, the division made by LZ algorithm is: γ, a, b, ab, abc d, c, bd, ab.

Instead of storing these patterns as a collection of substrings, we can build a tree containing them together with the number of times we can find each detected pattern by itself, or as a prefix of other patterns. This tree, called LZ tree, grows dynamically as the movement history is being processed. Figure 9.2 shows the LZ tree built when processing L. For instance, the node $[b:3]$ corresponds to pattern ab, which appears twice as ab (the pattern itself) and once as a prefix of another pattern (abc).

In order to make it easier to understand the parsing process and the evolution of the tree, Table 9.1 shows in the first row the sequence of locations recorded in L and in the second row the LZ parsing (i.e. the patterns detected and at which point of L they are recognized), which matches with the tree nodes added or updated at each step.

Once the algorithm processes each new symbol and then updates the tree, the next step is to calculate the probability for each symbol to be the one representing the most probable next location. For that purpose, LZ algorithm takes an approach proposed by Vitter [9], which calculation is made as shown by formula 9.1:

$$P(X_{n+1} = a) = \frac{N^{LZ}(la, L)}{N^{LZ}(l, L)} \tag{9.1}$$

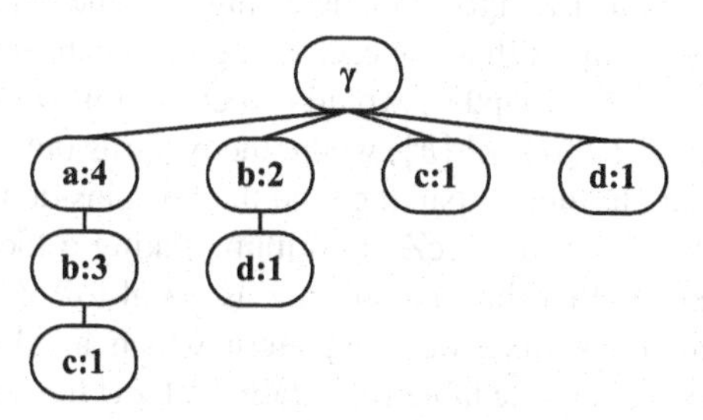

Fig. 9.2 LZ tree resulting from parsing the movement trace L

Table 9.1 Comparison of movement history parsing done by each LZ family algorithm

L	**a**	**b**	**a**	**b**	**a**	**b**	**c**	**d**	**c**	**b**	**d**	**a**	**b**
LZ	a	b	a	ab	a	ab	abc	d	c	b	bd	a	ab
LZU	a	b	a	ab	a	ab	abc	d	c	b	bd	a	ab
				b		b	bc				d		b
							c						
Win	a	b	a	ab	ba	ab	abc	bcd	cdc	dcb	cbd	bda	dab
ALZ	a	b	a	ab	ba	ab	abc	bcd	cdc	dcb	cbd	bda	dab
				b	a	b	bc	cd	dc	cb	bd	da	ab
							c	d	c	b	d	a	dab

where: $P(X_{n+1} = a)$ is the probability of symbol a being the corresponding to the most probable next location; l is called prediction context and stores the last substring parsed by LZ algorithm (e.g. we can see in Table 9.1 that $l = ab$ at step 6 whilst $l = a$ at step 12); $N^{LZ}(la, L)$ is the number of times the prediction context has been followed by symbol a in the LZ tree; and $N^{LZ}(l, L)$ is the frequency of pattern l in the LZ tree. This formula is calculated with all the different symbols detected up until that moment by changing a by the symbol about which we want to know the associated probability. Finally, we just take the location which symbol has achieved the highest probability.

This algorithm has three drawbacks: (i) patterns between two parsed patterns are not detected (in the example, *dc* is followed by *b*, but *cb* is not in LZ tree); (ii) patterns contained within parsed patterns are not detected either (*abc* pattern is in LZ tree, but *bc* is not); and (iii) Vitter method cannot make any prediction when a pattern is detected for the first time (if l is detected for the first time, it is not followed by any symbol, then $N^{LZ}(la, L) = 0$ for all symbols). The two next algorithms try to solve these problems.

9.2.2.2 LeZi Update Algorithm

The first variation over the basic LZ algorithm is intended to detect new patterns within the ones recognized by LZ algorithm. In order to do so, LeZi Update [1] makes the same parsing as LZ algorithm, detecting both the patterns resulting from LZ parsing as well as all the suffixes of each recognized pattern. Taking the former mobility trace example, LeZi Update would detect the following patterns: γ, a, b, $ab\{b\}$, $abc\{bc, c\}$, d, c, $bd\{d\}$, $ab\{b\}$, where the patterns outside the brackets are the ones corresponding to the LZ parsing, and the ones inside the brackets are the new ones detected by the Active LeZi algorithm. Taking a look at LZU tree row in Table 9.1, we can see that the first sub-row is equal to LZ row, whilst second and third rows contain the suffixes of each pattern, which are also added to the tree. Following the steps shown in the table we obtain the LZU tree in Figure 9.3.

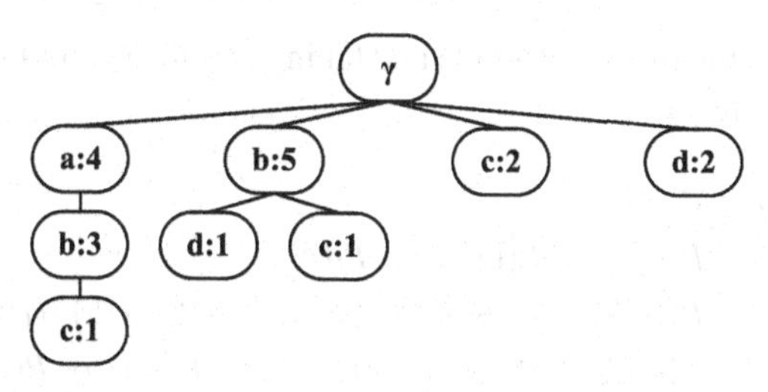

Fig. 9.3 LZU tree resulting from parsing the movement trace L

As in LZ case, once the predictor has processed the new symbol and updated the corresponding tree, it is time for calculating probabilities. LeZi Update bases this calculation on Prediction by Partial Matching (PPM) method [2], which solves the problem with Vitter method and takes into account more patterns in order to obtain the probabilities.

The probability calculation works as follows. We need to know firstly the longest prediction context, l, at each moment, i.e. the longest substring at the and of L which is already present in the LZU tree. Considering the former example, L, the prediction context is $l = ab$, since no substring dab is stored in the LZU tree. Once l is known, a table similar to Table 9.2 is built for storing the number of times that each detected substring has followed the longest prediction context (order 2 in our case, $l_2 = ab$) and the prediction contexts of lower orders up until order 0 ($l_1 = b$, $l_0 = \gamma$). Besides this information, the table includes also what is known as escape event, which is the number of times a pattern is not followed by any symbol. For instance, the pattern ab has a frequency equal to 3 but it has a child whose frequency sums 1 event, therefore there are 2 escape events. The last detail of the probability calculation method is related to the depth of the location prediction we want to obtain. As we are just focused on the next location, the probability calculation method is going to consider only one-symbol substrings following each prediction context. For instance, taking a look at the substrings following order 0 context, l_0, the probability calculation method is going to exclude substrings ab, abc, bc and bd as they are made up of more than one symbol.

Table 9.2 Frequency of the substrings following the current context for LeZi Update

$l_2 = ab$	$l_1 = b$		$l_0 = \gamma$				
c:1	c:1	esc:3	a:1	abc:1	bc:1	c:2	esc:0
esc:2	d:1		ab:2	b:3	bd:1	d:2	

Once the table is filled up and the calculation method knows what substrings to consider, the probability is calculated as shown in expression 9.2:

$$P(X_{n+1} = a) = P_k(a) = P(a|l_k) + P(esc|l_k) \cdot P_{k-1}(a) \tag{9.2}$$

which translated to the former example and taking c as the symbol we want to know its probability, results in:

$$\begin{aligned} P(X_{n+1} = c) = P_2(c) = & \\ & P(c|ab) + P(esc|ab) \cdot P_1(c) = \\ & P(c|ab) + P(esc|ab) \cdot \{P(c|b) + P(esc|b) \cdot P_0(c)\} = \\ & P(c|ab) + P(esc|ab) \cdot \{P(c|b) + P(esc|b) \cdot P(c|\gamma)\} \end{aligned} \tag{9.3}$$

and applying the data in Table 9.2:

$$P(X_{n+1} = c) = \tfrac{1}{3} + \tfrac{2}{3} \cdot \{\tfrac{1}{5} + \tfrac{3}{5} \cdot \tfrac{2}{13}\} \tag{9.4}$$

9.2.2.3 Active LeZi Algorithm

The last algorithm belonging to LZ family is Active LeZi [4]. This predictor has several modifications with respect to the two previous ones in order to add the patterns placed among consecutive patterns parsed by LZ algorithm. To achieve this goal, Active LeZi makes use of a dynamic window, which length depends on the longest pattern parsed by LZ algorithm after each new symbol is processed. The symbol processing made by Active LeZi algorithm is as follows. Whenever there is a new location (i.e. symbol), this predictor makes the same parsing than LZ algorithm, updating the length of the window when a new pattern is detected. Then, the so called ALZ tree is updated by adding the substring contained in the window as well as all its suffixes. The rows corresponding to the ALZ tree in Table 9.1 and the row corresponding to the window of this algorithm shows the evolution of the window as well as the nodes added to the tree after each new symbol is processed. Following the evolution shown in Table 9.1 we can get the ALZ tree of Figure 9.4.

The probability calculation method in this case is also based on PPM algorithm, but with some slight variations. Although expression 9.2 still applies when calculating the probability associated to each symbol, and the prediction contexts of different orders are selected in the same way, the table with the substrings following these contexts is different. In fact, it is different because we do not account for substrings, but just for the symbols following the contexts. Thus the resulting table for the former example in this case is the one shown in Table 9.3.

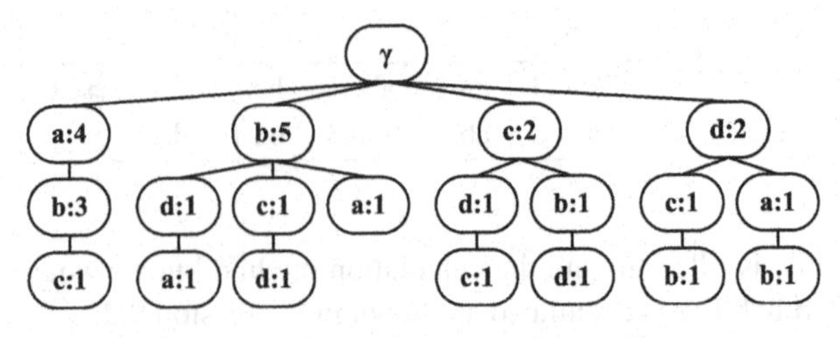

Fig. 9.4 ALZ tree resulting from parsing the movement trace L

Table 9.3 Frequency of the symbols following the current context for Active LeZi

$l_2 = ab$	$l_1 = b$		$l_0 = \gamma$		
c:1	a:1	d:1	a:4	c:2	esc:0
esc:2	c:1	esc:2	b:5	d:2	

With the variations introduced by this predictor, all the problems related to LZ algorithm are solved. However this improvement is not for free: it leads to a higher amount of information to be stored (we can see that ALZ tree of Figure 9.4 has many more nodes than LZ tree of Figure 9.2) and the time needed for calculating the probabilities is also increased since PPM algorithm is more complex than Vitter approach.

9.2.3 Performance Results

Leaving the theory aside, we present here some results of applying these algorithms on a set of traces of 95 different users, whose mobility has been tracked during an entire academic year [3].

We just show here the hit rate measurement (number of predicted next cell and actual next cell matches divided by the total number of processed cells after processing the whole trace). Figure 9.5 represents the percentage of traces (users) achieving, at least, the corresponding averaged (over the entire trace) hit rate. We can see that both LeZi Update and Active LeZi achieve a higher hit rate (more percentage of users have a higher hit rate) than when considering LZ predictor. More results and comparisons among the three algorithms in terms of hit rate and also resource consumption can be found in [6].

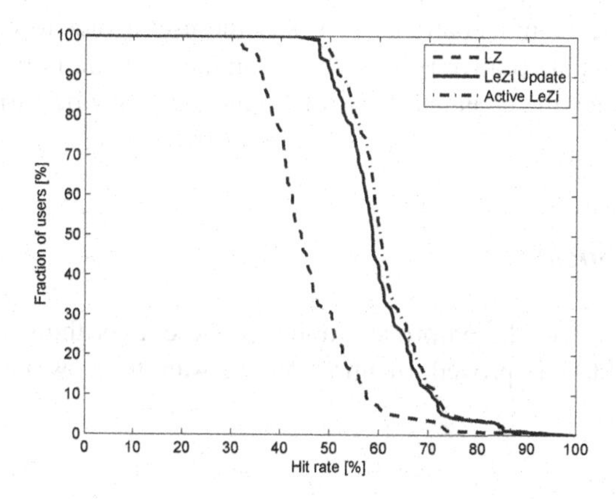

Fig. 9.5 Comparison of the percentage of users attaining, at least, the corresponding averaged hit rate for the three different LZ algorithms

9.3 Improving Recommender Predictions

As we have seen in the previous section, by using location prediction algorithms, a location-aware recommender in a mobile phone can give recommendations based not just on the current location, but also on future ones. However, the location prediction algorithms explained before have two main limitations. First, they cannot make any prediction when they do not have a history of past movement events to learn from. In other words, when a user visits a place for the first time, the predictors cannot make reliable predictions, they can just learn to predict in the future. This is a serious limitation for applications such as recommender systems for tourists. The second drawback is that these algorithms learn from previous behavior of the user, but they do not benefit from knowledge obtained from the rest of the users. In the following subsections we will face these two limitations.

9.3.1 Implanting "False Memories"

When a user visits a place for the first time, the movement trees managed by the prediction algorithms presented in Section 9.2 do not contain any information about that place, so they cannot make useful predictions. This would be a problem in a tourist guide application based on location-aware recommendations. To cope with this limitation, we propose grafting new branches in the movement trees, which would represent a "false memory"[1] of previous movements, that in fact corresponds to recommended routes for tourists. This "false memory" will cause the prediction algorithm to adapt to usual tourist routes, although the tourist herself has never been there before.

Therefore we could, for instance, create our own tree by designing its branches with the recommended paths. Then the user can download to her mobile phone the trees she is interested in (touristic guide of certain city, best restaurants or pubs...), and the algorithms will use that tree combined with the user's current context in order to make the recommendations. This way the user avoids to have fixed trips, because if she takes alternative path to the recommended one in that moment, the algorithm will notice the change and will recommend a new best trip in that same moment.

9.3.2 Going Social

Although the recommendation process done by these algorithms is dynamical in some way, it might be improved. Imagine that we want to know the best pubs of a

[1] This name is taken from the Total Recall movie, where memories of never visited places are implanted in the brain of a virtual tourist.

city according to our current location. Using a static tree that has some predefined branches may be a good solution for the day or week you retrieve that tree. However in cases like this one, the concept of "best" changes from one week to another. How to update the trees dynamically and fast, reflecting on them the changing "best" places? Since in some cases this "best" concept is closely related to the amount of people stopping by those places, it may be a good idea to use the collective intelligence. Therefore, instead of building the tree branches "by hand", why do not we let the users to create their social tree? Recalling ALZ tree, we saw that it contains all the existing patterns (as parsed by LZ algorithms). Thus, we can sum up the trees of the users by summing up the frequencies of the common nodes, and we will get some branches which weight (frequency of their nodes) will be bigger than the rest. Those branches will be the corresponding ones to the "best" paths. Moreover, since each user's tree changes as the user moves and visits new places, the social tree also changes and adds those new places as well as the changes on the most visited ones. Therefore in order to implement the social version of the recommendation system each user would have two trees: one of them being updated as explained in the previous section in order to track and learn the individual patterns (and thus, most visited paths) and sent to a central database so as to add it up to the rest of trees; and the second tree will be the social tree, downloaded each time the user wants to run the recommendation system, and used for searching recommendations based on the current context.

Up till now we have explained theoretically how location prediction algorithms could be useful in order to make recommendation systems focused on suggesting places to visit as the user moves. To complete the idea, we are going to describe two possible examples to which this recommendation system could be applied, at two levels: social but static, and social as well as dynamic.

The first example is a touristic guide. We have labeled it as a social but static recommendation scenario because, although the most relevant or interesting places usually are the ones most visited, the touristic places do not change very often. Therefore it would be needed to add the users' trees up until reaching a stable tree. After that, the changes in the tree structure (and therefore, in the interesting paths to follow) would be insignificant. However, the recommendation system would still preserve the dynamical feature of recommending touristic paths as the user moves and decides whether to follow the recommended path or search for more interesting places by ignoring the last recommendation and following another path. In that moment the system would notice the user's decision and suggest the best path based on the new current context.

The social and dynamical example is the one involving shops, restaurants, pubs and similar places, whose popularity changes very fast. In this case, the idea of a social tree that changes at the same speed than the users' preferences fits perfectly when building up the recommendation system.

9.4 Prototype Implementation

To demonstrate the feasibility of implementing LZ-based location prediction algorithms to refine the suggestions presented to the user by a location-aware recommender in a mobile phone, in the context of the España Virtual project [2] we have developed a prototype for Android OS [5] that has been tested on a mobile phone HTC Desire with Android version 2.2.

In this prototype we have used the information provided by the GSM/UMTS base stations, the cell identifier (CellId) and location area code (LAC), to obtain the current location of the mobile phone user with an accuracy that ranges from 50 meters (in urban areas) up to several thousand meters (in rural areas). We decided not to use GPS due to its high power consumption and coverage problems indoors.

When the application detects a cell change, it gets the information (CellId and LAC) of the new cell and obtains the coordinates (latitude and longitude) of the current location using the Google Geolocation API [3]. To avoid repeated queries to this API and save future network traffic, the coordinates of already visited cells are stored in a cache in the mobile phone. Once obtained the coordinates, the CellID and LAC are concatenated into a single identifier and passed to the three prediction algorithms presented in Section 9.2 (Active LeZi, LZ, and LeZi Update) and a prediction of the two most probable next locations are obtained using these algorithms.

Then, using the latitude and longitude coordinates, the application uses the Google Places API [4] to obtain information about the establishments or points of interest around that location. The interaction with this API is made in two steps. First, a Place Search request initiates a request for "places" around a provided location, and returns a list of candidate places that are near the provided location. Then, the application initiates a Place Details request on each specific candidate place returned by Place Search to obtain more information about each particular establishment or point of interest. This process is repeated for the coordinates of the current location, and for the ones obtained as most probable future locations from the location prediction algorithms, and the results are presented in real-time to the user. Then, the application waits for future cell changes, and the process is repeated until the application is finished. In figure 9.6 we show a screen capture of the prototype application.

In the current prototype, the recommendations are filtered only by location, and no other criteria (user profile, etc.) is used. The probability of future locations obtained from the algorithms is also presented in the screen, although it is not used for filtering the results presented to the user.

We have found an unexpected practical problem during the implementation of the prototype. It was a limitation of the operating system, that affected the way we had to implement the application. Android API provides a component, called Service, for

[2] `http://www.espa~navitual.org`

[3] `http://code.google.com/intl/es-ES/apis/gears/api_geolocation.html`

[4] `http://code.google.com/intl/en/apis/maps/documentation/places/`

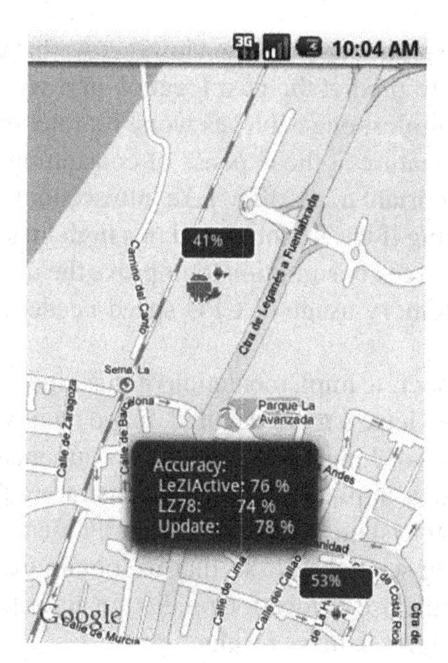

Fig. 9.6 Screen capture of the prototype application

doing background processing. A Service is an application component that remains running even when the application is not on foreground (as opposite to Activity components, which might be killed by the OS without the user knowledge when the application is not on foreground). So far it seems that a Service is just what we need for monitoring cell changes. However there is a limitation. When an Android device is idle for a while, it automatically enters a power saving mode because of which CPU is turned off. Therefore all the tasks, included the background cell change monitoring, are stopped. There is a class, called `PowerManager.Wakelock` (from PowerManager API) that allows for an application to take control over this power consumption saving mode, keeping the CPU running when needed, or leaving even the screen on or dimmed (not full bright) as long as the application needs. As one may expect, this control over the power consumption saving mode is not for free; it leads to an increase of the battery drain. Therefore further improvement of this application is needed in order to be used for final users in real scenarios.

9.5 Conclusions

Along this chapter we have seen some location prediction tools that can be used in order to implement systems for recommending places the user may be interested in as she moves. Taking a step further we have also described, broadly speaking, how the translation from prediction to recommendation could be done as well as some applications for which the resulting recommendation systems would be useful.

Regarding the prediction algorithms covered in this chapter, we have focused on LZ family since they can be executed directly on mobile devices, which leads to the many advantages we mentioned before. However we have shown the significant differences when trying to predict the next location of a wide set of users with different algorithms: the simplest one achieves worst hit rate whilst the most complex predictor improves this feature at the expense of consuming much more resources. Therefore there is an important trade-off to take into account when considering what predictor to use, depending if we are interested in a high hit rate at any cost, or if we prefer to reduce the resource consumption to improve the user experience (reducing battery consumption, memory usage or CPU speed needed) by assuming a higher probability of error.

Nevertheless some concrete implementation of this kind of recommendation systems is needed in order to have a more detailed image on how these location predictors really perform when working as the core of recommendation systems. We have shown a prototype of such implementation, demonstrating this way the feasibility of building recommender systems based on LZ family location predictors, but there is still a need for making several tests so that we obtain extensive data to analyze the improvement due to the use of location prediction algorithms. Hopefully we will not have to wait too long to see these results.

Acknowledgements. This work has been partially supported by the España Virtual project, led by DEIMOS Space and funded by CDTI as part of the Ingenio 2010 program (Spanish Ministry of Science and Innovation). It has been also partially supported by the Spanish Ministry of Science and Innovation through the CONSEQUENCE project (TEC2010-20572-C02-01).

References

1. Bhattacharya, A., Das, S.: LeZi-update: an information-theoretic framework for personal mobility tracking in PCS networks. ACM/Kluwer Wireless Networks J. 8(2-3), 121–135 (2002)
2. Cleary, J., Teahan, W.: Unbounded Length Contexts for PPM. In: Proceedings of the Data Compression Conference, DCC 1995, pp. 52–61 (1997)
3. Eagle, N., Pentland, A., Lazer, D.: Inferring Social Network Structure using Mobile Phone Data. Proceedings of the National Academy of Sciences (PNAS) 106(36), 15274–15278 (2009)
4. Gopalratnam, K., Cook, D.: Online sequential prediction via incremental parsing: the Active LeZi algorithm. IEEE Intell. Syst. 22(1), 52–58 (2007)
5. Meier, R.: Professional Android Application Development. Wrox Press Ltd., Birmingham (2008)
6. Rodriguez-Carrion, A., Garcia-Rubio, C., Campo, C.: Performance Evaluation of LZ-based Location Prediction Algorithms in Cellular Networks. IEEE Commun. Lett. 14(8), 707–709 (2010)
7. Song, L., Kotz, D., Jain, R., He, X.: Evaluating Next-Cell Predictors with Extensive Wi-Fi Mobility Data. IEEE Trans. Mobile Comput. 5(12), 1633–1649 (2006)
8. van Setten, M., Pokraev, S., Koolwaaij, J.: Context-Aware Recommendations in the Mobile Tourist Application COMPASS. In: De Bra, P.M.E., Nejdl, W. (eds.) AH 2004. LNCS, vol. 3137, pp. 235–244. Springer, Heidelberg (2004)

9. Vitter, J., Krishnan, P.: Optimal prefetching via data compression. Journal of the ACM 43(5), 771–793 (1996)
10. Yang, W.S., Cheng, H.C., Dia, J.B.: A Location-Aware Recommender System for Mobile Shopping Environments. Expert Syst. Appl. 34(1), 437–445 (2008)
11. Yap, G.E., Tan, A.H., Pang, H.H.: Discovering and Exploiting Causal Dependencies for Robust Mobile Context-Aware Recommenders. IEEE Trans. Knowl. Data Eng. 19(7), 977–992 (2007)
12. Ziv, J., Lempel, A.: Compression of individual sequences via variable-rate coding. IEEE Trans. Inf. Theory 24(5), 530–536 (1978)

Chapter 10
SCORM and Social Recommendation: A Web 2.0 Approach to E-learning

Rebeca P. Díaz Redondo, Ana Fernández Vilas, José J. Pazos Arias, Alberto Gil Solla, Manuel Ramos Cabrer, and Jorge García Duque

Abstract. This paper introduces the *Educateca* project, a Web 2.0 approach to e-learning. The project refactors SCORM, the *de facto* e-learning standard, to embrace the two main shifts in Web 2.0: the WOA (Web Oriented Architecture) and the social trends in user involvement. Thus, our proposal combines (i) a more dynamic strategy where brief pedagogical units are offered as services to be combined on-the-fly with (ii) a more active role of students, who become also content producers instead of being mere consumers. In this context, content recommendation plays an essential role to avoid overwhelming users with too much educative content that are not able to filter, asses and/or consume. Along this line, we introduce in this paper a new recommendation mechanism supported by the tag clouds which label both users and content.

10.1 Introduction

E-learning and its related technologies and standards are destined to accept the two main paradigm shifts in the Web 2.0 era [1, 2]: the programmable Web and the social Web. In the programmable Web vision, the topology behind WOA (Web oriented Architecture) makes invalid the typical scenario where an organization stores learning materials locally in an LCMS (Learning Content Management System) and deliver them through an LMS (Learning Management System) to the learner's device. Instead, learning organizations should publish and even share their material in the cloud so that learners can directly access them. In the same way, the LMS is becoming another SaaS (Software as a Service) in the cloud, so that students can select a LMS according to the LMS's features and their preferences. To sum up, learners, content and LMSs are not necessarily in the same domains or from the

Rebeca P. Díaz Redondo · Ana Fernández Vilas · José J. Pazos Arias · Alberto Gil Solla · Manuel Ramos Cabrer · Jorge García Duque
SSI Group, Department of Telematic Engineering, University of Vigo, 36301 Vigo, Spain
e-mail: {avilas,jose,rebeca}@det.uvigo.es

J.J. Pazos Arias et al.: Recommender Systems for the Social Web, ISRL 32, pp. 195–207.
springerlink.com

same providers in a WOA. As the different elements move towards service-based approaches, consumers (learners in our case) may opt to mix and match a variety of tools from multiple providers to build an environment that will work best for them.

Apart from these architectonic changes, the Web 2.0 era is characterized by a strong social component. The cloud has to be a place where the learning content is easily generated and published by both teachers and learners, so that the collective intelligence of users encourages a more democratic use. What is more, users may tag, comment and rate any pedagogical content. These opinions should be used to recommend and filter content with the aim of offering the more convenient units to each student. This is the emphasis of this paper, introducing the recommendation strategy for learning services in *Educateca* project.

The *Educateca* project is precisely based on these Web 2.0 social and technological basis and, in order to provide a interoperable environment, it embraces the ADL SCORM [3] (Shareable Content Object Reference Model): the *de facto* standard in the learning field. The paper is organized as follows. In the next section we briefly introduce the changes we propose in the SCORM to properly accomplish both social and WOA aspects of the Web 2.0. Section 3 overviews the *Educateca* vision in the new context of the cloud and in Section 4 we detail the algorithms we propose to provide collaborative recommendation for learning content. Finally, conclusions and future work is exposed in Section 5, as well as a comparison between our approach and other works about bringing Web 2.0 to e-learning.

10.2 SCORM Shifts to Web 2.0

Although SCORM encourages the reusability, some aspects are not ready to Web 2.0 immersion [4]. With the aim of tackling these shortcomings, *Educateca* project refactors SCORM infrastructure to provide e-learning in the Web 2.0 arena. In this section we introduce the changes we propose in this line.

10.2.1 Brief SCORM Overview

Although SCORM is a vast standard that deals with any aspect related to any e-learning aspect, we only focus on SCORM CAM (Content Aggregation Model) and SCORM RTE (Run-Time Environment), the two most relevant components for our approach.

On the one hand, the SCORM CAM defines three main elements used to build a learning experience from learning resources: (i) SCO (Shareable Content Object), a single launchable learning object that utilizes SCORM RTE to communicate with a LMS; (ii) Content Organization, a map that represents the intended use of the content through structured units of instruction; and (iii) Metadata that should be used to describe these elements in a consistent manner; SCORM strongly recommends the use of the IEEE LOM (see Figure 1). Besides, Content Aggregation is an entity used to deliver both the structure and the resources that belong to a course in a Content Package, which consists of a compressed file with the physical resources of educational content and at least one XML file –called manifest– that embodies a structured inventory of the content of the package: its organization (<organizations>) and metadata (<metadata>).

On the other hand, SCORM RTE defines: (i) the launch process as a common way for LMSs to start SCOs; (ii) the API as the mechanism for exchanging data between the LMS and the SCO; and (iii) a data model as a standard set of elements to define the information tracked for a SCO.

10.2.2 Architectonic Shift: Driving SCORM to the WOA

The concept of cloud computing arose as consequence of both the progressive price reduction of consumer electronic devices with computing capabilities and the growing availability of broadband connectivity. Although in origin the cloud was conceived as a way to allow users to access their documents and applications from anywhere and using any device as a lightweight client, this vision has expanded to embrace the so-called Everything as a Service (EaaS or XaaS) trend. Thus, in addition to applications for on-demand use (Software as a Service), the offer encompasses computation, storage, development and communication resources (Platform as a Service, Communication as a Service) and, at the same time, it integrates whichever services may be provisioned through the e-commerce technologies, including human resources in any areas of activity (what we may term User as a Service). Therefore, the cloud lodges arbitrary resources that appear as high-granularity services that may be composed in a flexible manner in response to complex necessities. In this new context, both learning software and content should not be delivered in a packetized way anymore. Instead, we propose to offer these elements as autonomous ones with the aim of being available to be used separately according to the users' needs.

Therefore, to integrate SCORM content and processes in the WOA cloud we propose the following modifications. First, the compressed file structure of a SCORM course (learning content) would be replaced by set of independent pedagogical units able to communicate to the SCORM RTE to be run (independently or cooperatively) whenever needed. Thus, each SCO becomes an SCS (Sharable Content Service) according to the SaaS conception. Besides, the organization of a SCORM course remains, but it is separated from the learning objects itself. This provides more flexibility to users which finally decide if they prefer a pre-established course definition or an ad-hoc composition of pedagogical units according their interests.

Secondly, the SCORM RTE itself (learning software) is deployed as a set of Web Services offering the SCORM API. Finally, the SCORM CAM (Content Aggregation Model) is replaced by a SAM (Service Aggregation Model), a service (learning

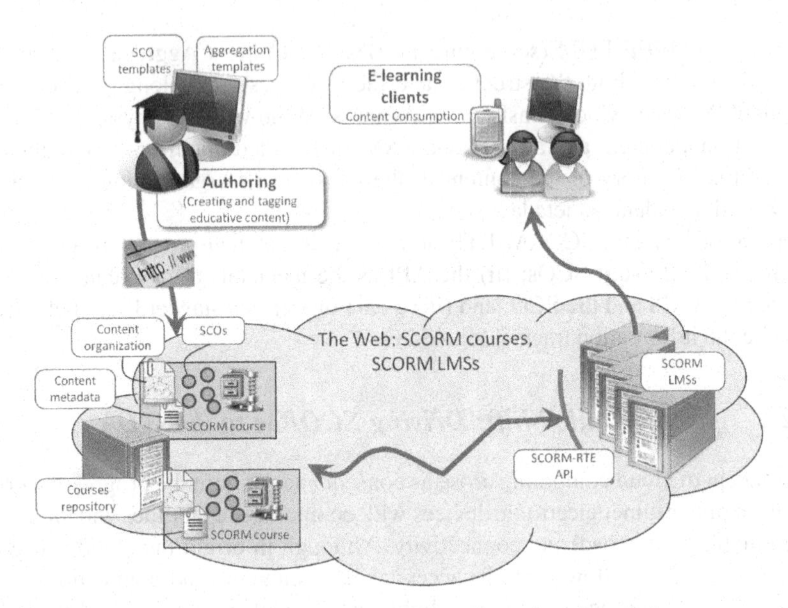

Fig. 10.1 Typical architecture for any SCORM-compliant e-learning platform

software) in the cloud that provides access to a set of SCSs according to a specific organization.

10.2.3 Social Shift: Collaborative SCORM

Social networks appear as a true social and cultural phenomenon on the Internet, whose impact is evident in any social sector. In fact, there are millions of users who already engage in online social interactions as part of their daily lives: sharing information, exchanging experiences and opinions, tagging elements, ratting content and services, even creating or uploading their own content to be shared to other users. Thus, social networks have facilitated a notable change of attitude in users: from a more passive to a really active on. This change is not only restricted to the social networks context, it has spread out to any aspect in the Internet, and e-learning cannot stand aside.

Consequently, we propose an environment where learners have a more leading role in a context where they are no longer passive receivers of information given by a teacher, but active participants whose opinions are took into account. Firstly, describing and cataloguing learning objects is no more responsible for the teaching committee. On the contrary, learners tag and rate learning objects as they use the platform. Secondly, not only tagging but authoring also moves from teachers to learners so that these can create or modify content to contribute to the learning distributed repository. Besides, users can create and share course organizations combining pedagogical units created by them or by any other user.

This vision of providing a suitable space for active learners comes up against the SCORM conception. SCORM was created assuming a single authority (usually a teacher) that creates and assigns metadata for a collective of students. This metadata theoretically allows searching and combining learning objects to make them accessible for other purposes, but, in practice, SCORM content hardly include metadata. Consequently, it is difficult for a community to alter or add metadata tags within SCORM infrastructure, and it is currently impossible for a learner to contribute to a course by adding or altering its content.

Therefore, we propose to change the processes which affect metadata and organization specification to support this environment of participation with SCORM-compatible infrastructures. Apart from distributing authoring (both of learning objects and organizations), a Web 2.0 approach requires an open access from students to metadata so that both learning objects and organizations are collaboratively tagged.

10.3 EDUCATECA Overview

Figure 2 depicts the *Educateca* interpretation of Web 2.0 principles in the e-learning field. Being inspired by our personalized e-learning platform T-MAESTRO [5], we have added social and WOA conceptions, central milestones of the Web 2.0. One of the aspects that must be highlighted is how teacher and student roles change in this new model. Whereas in the traditional model (Figure 1) both roles are totally separated, in this new approach students are expected to adopt a prosumer attitude. Thus, students are not more those who receive courses and pedagogical units and whose only obligation is learning the provided content and demonstrate their new knowledge resolving an exam; instead they have now the opportunity of providing new educative content.

According to the WOA interpretation of a SCORM course, their two main components are the SAM (Service Aggregation Model) and SCS (Sharable Content Service), respectively mirrors of CAM and SCO. The CAM service maintains the structure and location of the SCSs in a course but not the SCSs themselves; whereas a SCS consists mainly of an XML template of the learning content, the metadata describing that content and the location of the resources (assets) in the template. Consequently, the creation of new educative contents can be done by prosumers in two ways: (i) creating their own pedagogical units (SSCs) and (ii) defining new service aggregations that combine already developed SSCs.

To create new pedagogical units, prosumers may use any SCORM-complaint authoring tool that is offered as a service in the cloud. To define new service aggregations we propose using BPEL [9], which provides a standard mechanism for services aggregation according to certain rules. In our case, these rules are the sequencing rules defined in the SCORM 2004 Sequencing and Navigation specification: rules responsible for determining what happens when learner exits a SCO.

With this aim, we have translated the SCORM terms into BPEL tags, which must be understood by the LMS to provide the desired sequence of events. Thus, it is necessary to implement a set of algorithms that apply the sequencing rules to the current set of tracking data to determine which activity should be delivered next. *Educateca* incorporates these algorithms in the LMSs by integrating the ActiveBPEL Engine (`http://www.activevos.com`). Since every user in the community should be able to publish a new course organization or a modified version of an existing one, *Educateca* sites provides SAM templates to make it easier.

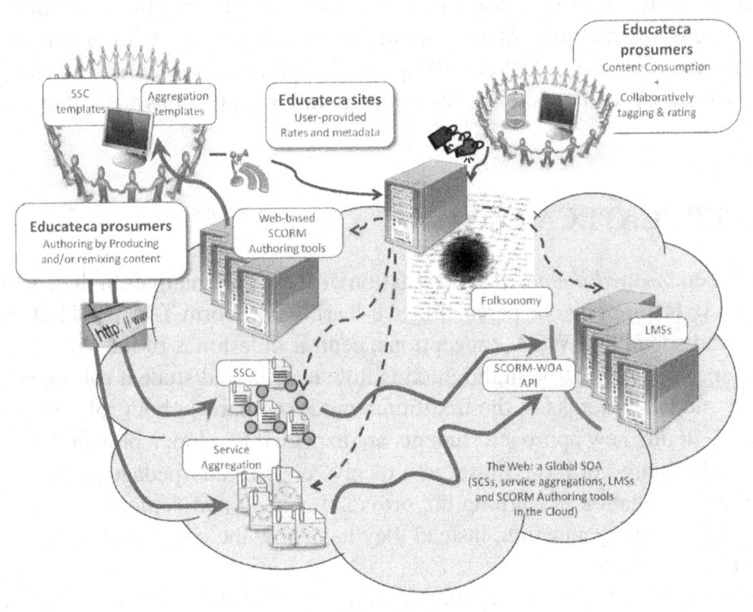

Fig. 10.2 Proposed architecture for SCORM-compliant e-learning 2.0

The active role of students goes further than creating new elements: prosumers have also the possibility of given their opinion about other SSCs and service aggregations as well as labeling them. Tags given by prosumers allow creating and dynamic updating a folksonomy which support a mechanism to classify and search educative units whenever needed. In fact, users look for pedagogical content according to key-words and their decision of choosing one or another will be likely based on the rates each unit has received. So, instead of having an environment where the quality of content is assumed (like in classical models), this new model entails a certain degree of competition where the opinion of consumers gives a measurement of the quality. Finally, and for courses consumption, users may use any SCORM-complaint LMS having a Web-based interface, just according to the main SCORM principles that claims for open e-learning environments.

E-learning in the cloud assumes a model where a huge amount of LMSs having different features (cost, efficiency, level of personalization, etc) would be available. In our approach, selecting an appropriate LMS is an *Educateca* site's task. To support this selection, we propose the *Educateca* provider and the student define a SLA (Service Level Agreement). Concretely, *Educateca* sites virtualize LMS by offering a Virtual Web Service, in the sense defined in our previous work in [6, 7], which we call VLMS (Virtual-LMS). Client Device uses the V-LMS as any other service in the cloud and only *Educateca* sites are aware of the virtual nature of the service and materialize the service in a real LMS by incorporating user-defined preferences, load-balancing, availability, caching, etc.

10.4 Collaborative Support to Recommendation

In order to support the e-learning community, *Educateca* sites incorporate mechanisms to assist the users in a personalized access to e-learning material (discovering and recommendation). To be precise, *Educateca* strategy is based on collaborative tagging. Folksonomy-like structures are maintained for users and e-learning material in the cloud, which support social collaborative filtering to discover and recommend e-learning material. The *Educateca* community collaborates in creating, sharing and describing content so that students and teachers may tag not only their own content but also the other's content. Each SCS, SAM or even LMS is associated with a set of weighted tags (tag cloud). Similarly, the users in *Educateca* community have their own tag cloud, representing his/her profile. The details of the recommendation strategy are described in the following sections.

To establish a solution for recommending learning material in a social environment, we need an automated model of both the learning resources (services in our case) and the behavior of the learners regarding these resources. From the tagging perspective, users assign tags to learning material in the system. However, the utility of the system is not tagging resources but assisting the user in order to find learning services. That is the filtering perspective. If the tagging activities turn into collaborative in the social era, the filtering techniques are socialized as well. The socialization of recommendation has two sides. On the one hand, collaborative filtering, taking into account other like-minded users (neighbors), different from the one for who recommendation is being computed, turns into the natural scheme. On the other hand, the filtering mechanisms should incorporate some trust-based criterion for selecting recommendable resources. This trust-based criterion should be combined with the collaborative filtering one in order to select the users whose opinions are taken into account for recommendation.

The emphasis of the paper is introducing the tagging and the utility model for social consumption in *Educateca* recommendation: a recommendation technique which combines collaborative tagging and collaborative filtering. The tagging model is a folksonomic structure which describes learning material. Over this structure, the utility model for recommendation is constructed by using tag clouds. This technique can be combined with a trust-based criterion in order to obtain a recommendation schema when a social structure among users is available: collaborative tagging and trust-based relationships among users.

10.4.1 Modeling Learning Services as Perceived by Tagging Users

The social vocabulary for describing services is a folksonomic structure which mathematically can be described [16] as a tuple $F = (U,T,R,Y)$ where $U = (u_1,\ldots,u_m)$, $T = (t_1,\ldots,t_l)$, and $R = (r_1,\ldots,r_k)$, are finite sets of resources, tags and users (who assigns tags). $Y \subseteq UxTxR$ is a ternary relation whose elements are called tag assignments; tags which describe the learning services (resources) as freely perceived by the users.

According to Hotho notation, for describing the service, we have to pay attention to the TR Model (Tag-Resource Model), abstracting away the specific user who tags the service. This model considers tags in *Educateca* at the resource level, so that the set of tags given by all the users to a service –$T(s)$– is treated as a document; and the tags t as terms of this document. The weight of a tag t for the service s –$w(s,t)$– and the Tag Cloud for a service s, $TC(s)$, is computed as:

$$w(s,t) = \frac{tf(t,T(s))}{l(T(s)} \qquad TC(s) = \{t, w(s,t) | t \in T(s)\}$$

where tf and l are the term frequency and the length of the document respectively. Therefore, for recommendation purposes, a learning content is represented by a set of tags (graphically by a tag cloud) with their respective weights, which are proportional to the number of tagging users who have assigned this tag to the resource.

10.4.2 Modeling Learners

Until now, we have defined a model to describe the resources in a context of collaborative tagging. However, a recommendation engine requires establishing a preference model for users, also connected with the resource model, which record the preferences of the user regarding the utility of items.

Users in *Educateca* site can play the role of taggers when they tag the learning services or learners (service consumer) when they only use the learning services. The latter activity, consuming learning services, as the main form of learning, is the one which has to be monitored in order to recommend learning services. At this respect, a learner l is modeled by accumulating its services consumption. That is, the tags of the services the user consume $T(l)$ are aggregated in a weighted way to obtain a tag cloud which represent the learning interests. The weight of a tag t for the learner l –$w(l,t)$– and the Tag Cloud for a learner l –$TC(l)$– is computed as:

$$w(l,t) = \frac{tf(t,T(l))}{l(T(l)} \qquad TC(l) = \{t, w(l,t) | t \in T(l)\}$$

where, as in service model, tf and l are the term frequency and the length of the document $T(l)$.

10.4.3 Simple Commparison of Tag Clouds

In order to deploy a recommendation strategy which selects learning services suitable (according to past consumption habits) for a specific user, all the well-known techniques – content-based, collaborative, knowledge-based, etc.– need for some form of comparison between the background data (information the system has before the process starts) and the input data (information the user communicates to the system to generate the recommendation). In our case this data is always in the form of a set of tags. The Tag Clouds of learners and services is information the system has ; the information the user provides to the system can be also a set of tags in order to discover learning content. Besides, the concrete comparison depends on the type of recommendation technique. For instance, collaborative techniques need to find users similar to the target user in order to extrapolate the interest of a target resource; content-based techniques need to compute similarity between the target resource and resources which has been previously consumed by the target user; etc.

The simplest way to measure the similarity between two Tag Clouds, TC_i and TC_j, would be to count the number of coincident tags in both tag clouds, i.e., the higher the number of coincident tags, the higher the degree of relationship between TC_i and TC_j. One step further is taking into account the relative importance of the coincident tag ($w(s,t)$ or $w(l,t)$) in the tag cloud, so that the TC_i and TC_j are vectors in the tag space. Several techniques can be used which compute the similarity between vectors in the space of tags (Jaccard coefficient, Dice similarity, Cosine similarity) by considering that every resource (service in our case) is modeled as a vector over the set of tags, where each weight $w(s,t)$ in each dimension corresponds to the importance of a particular tag t. In *Educateca*, we have implemented cosine similarity so that the similarity between two Tag Clouds TC_i and TC_j is the cosine of the angle between them (determining whether the two Tag Clouds are pointing in roughly the same direction).

$$simcos_d(TC_i, TC_j) = \frac{\sum_{t_k \in TC_i \cap TC_j} w(TC_i, t_k) \cdot w(TC_j, t_k)}{\sqrt{\sum_{t_k \in TC_i} w^2(TC_i, t_k)} \sqrt{\sum_{t_k \in TC_j} w^2(TC_j, t_k)}}$$

10.4.4 Semantically-Enhanced Comarison of Tag Clouds

In the previous section, we derive a similarity measure between Tag Clouds which only takes into account the tag coincidence. That is, a direct Tag Cloud comparison which obviates the semantic which emerges from the collaborative tagging process. This direct approach does not take into account the potential semantic relation between the tags in the folksonomy. That is, although a tag t in TC_i does not belong to TC_j (or appears with a lower weight), it can be closely related (be very similar) to a tag t' in TC_j . To take this fact into account, we do not only consider the number of coincident tags and their weights (direct cosine similarity simcosd), but also the degree of relationship between the tags of both tag clouds. In this section we propose a semantically-enhanced similarity for Tag Clouds which computes the similarity by considering not a set of tags in the folksonomy, but a network of tags where there is a relationship between every two tags in the folksonomy. This structure formalizes the collective intelligence in *Educateca* and it is the basis for improving Tag Cloud comparison. Once again, the specific Tag Clouds to compare depend on the kind of recommendation strategy as it is shown in the following section.

Taken as a whole, the aggregation of many annotations (from many users) results in a complex network of interrelated tags, where r_{ij} represents the similarity relation between the tags t_i and t_j. Extracting this relation is precisely what we need in order to improve similarity measures in contexts of collaborative tagging. In a first approach, tags can be related in a way that the more times the tags are assigned together to the same resource, the highest the weight of their relationship:

$$r^d_{ij} = r^d(t_i,t_j) = \frac{\sum_{s \in S(t_i) \bigcap S(t_j)} w(s,t_i) \cdot w(s,t_j)}{|S(t_i) \bigcap S(t_j)|}$$

where $w(s,t)$ is the weight of the tag t in a Tag Cloud of the service s; and $S(t)$ is the set of the services in the system which has the tag t. On set further, we can improve r_{ij} by considering that two tags are similar if they appear in services which are similar:

$$r^i_{ij} = r^i(t_i,t_j) = \frac{\sum_{s \in S(t_i); s' \in S(t_j)} w(s,t_i) \cdot w(s',t_j) \cdot sim\cos_d(TC(s),TC(s'))}{|S(t_i) \times S(t_j)|}$$

Whether r^d_{ij}, r^i_{ij} or a combination, from the structure of interconnected tags, we define the indirect cosine similarity $sim\cos_i$:

$$sim\cos_i(TC_i,TC_j) = \frac{\sum_{t_k \in TC_i; t_l \in TC_j} w(TC_i,t_k) \cdot w(TC_j,TC_l) \cdot r_{kl}}{\sqrt{\sum_{t_k \in TC_i} w^2(TC_i,t_k)}\sqrt{\sum_{t_l \in TC_j} w^2(TC_j,t_l)}}$$

which computes the cosine by using a vector augmented with the relationships in the structure. That is the similarity between two Tag Clouds, TC_i and TC_j, has into account, not only the coincident tags and their weights, but also tags which are similar from the point of view of collaborative taggers. The indirect cosine similarity $sim\cos_i$ introduce the degree of relationship r_{ij} between the tags in both Tag Clouds. In this manner, the similarity between TC_i and TC_j, referred to as *FolkSim*, is defined as the weighted average of direct and indirect similarity where α is the adjustment parameter:

$$FolkSim(TC_i,TC_j) = \alpha \cdot sim\cos_d(TC_i,TC_j) + (1-\alpha) \cdot sim\cos_i(TC_i,TC_j)$$

10.4.5 Hybrid Collaborative Filtering Scheme

From the variety of techniques proposed to compute recommendation, in practice, the more successful strategies combine collaborative filtering with some other technique to gain better performance with fewer of the drawbacks of collaborative filtering. This is also the approach in *Educateca*. To be precise, we propose a hybridization of collaborative and content-based strategies which allows alleviating some of the problems of pure collaborative and pure content-based recommenders.

A content-based recommender for *Educateca*, would be based on the features of learning services, that is, the Tag Cloud of the services. This pure recommendation strategy can be expressed as follows.

$$Rec_{CB}(u,s) = \frac{1}{|S(u)|} \sum_{s' \in S(u)} Folksim(TC(s), TC(s'))$$

where $S(u)$ is the set of services which the user u has consumed. The idea is that a new service is recommendable to a user if it is very similar to the services consumed in the past. However, a pure content-based recommender would suffer the new user ramp-up problem, which is located in users which has not consumed any service (so that $S(u) = \{\}$), and the cross-genre problem or *"ouside-the-box"* recommendation ability –recommendation is computed only form the preferences of the user so that it tend to recommend very similar items, leading to a limited diversity in the recommendations. To alleviate these problems, collaborative recommenders generate recommendations based on inter-user comparison. That is, recommending content appealing to other like-minded viewers (named neighbors) so it provides much more diverse recommendations. Unfortunately, they only work well for users who fit into some stereotype with many neighbors of similar taste, that is, it does not work well for so-called "gray-sheep". In order to mitigate these problems, we combine a collaborative recommender with a content-based one in the following way:

$$Rec_{CF}(u,s) = \frac{1}{|Neighbors(u)|} \sum_{u' \in Neighbors(u)} Folksim(TC(u), TC(u')) \cdot Rec_{CB}(u', s)$$

where the users u' which are neighbors of a user u –$Neighbors(u)$– are determined by some threshold in $Folksim(TC(u), TC(u'))$.

10.5 Conclusions

There are some other approaches in the literature which transfers e-learning platforms to a service-oriented approach and even to cloud e-learning. In [10], Vossen & Westerkamp discuss typical problems of SCORM-related standards and propose a service-oriented approach as a solution, but any specific architecture or implementation is introduced. In [11, 12] Dong proposes the use of cloud computing as a base for modern e-learning by introducing the BlueSky cloud framework which virtualizes physical machines for e-learning systems. Similarly, CoudIA (Cloud Infrastructure and Application) project [13] runs private cloud infrastructure for e-learning and collaboration in the university environment of HFU (Hochschule Furtwangen University). With a lab-oriented application, Virtual Computing Laboratory (VCL) [14] (by North Carolina State University) enables students to reserve and access virtual machines (VMs) with a basic image or specific applications environments, such as Matlab and Autodesk. Meanwhile the above works focus on the infrastructure dimension of cloud-based e-learning, our proposal gives a more ambitious conception for e-learning in the cloud. Not only the infrastructure but the content and the metadata are freely in the cloud and then reassembled on demand by *Educateca* sites. Apart efrom that, the social dimension of e-learning in the cloud is considered by turning students and teachers into a peer-to-peer community.

The *Educateca* project is one more step towards new perspectives for enhanced learning with SCORM. The *Educateca* site and the online authoring tool are based on our previous work in t-MAESTRO [5], a SCORM-compatible infrastructure for personalized t-learning experiences combining TV programs and learning contents in a personalized way, with the aim of using the playful nature of TV to make learning more attractive and to engage TV viewers in learning. To be precise, we have adapted the functionality and architecture of the t-MAESTRO ITS (Intelligent Tutoring System) which constructs the t-learning experiences by applying semantic knowledge about the t-learners; and the A-SCORM Creator Tool, the authoring tool which allow teachers to create adaptive courses with a minimal technical background. Finally, the introduction of a folksonomy-based approach to provide a suitable recommendation mechanism was also based on our previous work [8, 15], although we have revised and introduced new capabilities to assess learning content according to the users' perception, which is definitively more precise in a context like the *Educateca* site.

Acknowledgements. This work has been partially funded by the Ministerio de Educacion y Ciencia (Gobierno de Espana) research project TIN2010-20797 (partly financed with FEDER funds), and by the Conselleria de Educacion e Ordenacion Universitaria (Xunta de Galicia) incentives file CN 2011/023 (partly financed with FEDER funds).

References

1. Chatti, M.A., Jarke, M., Frosch-Wilke, D.: The future of e-learning: a shift to knowledge networking and social software. International Journal of Knowledge and Learning 3(4/5) (2007)
2. Downes, S.: e-learning 2.0. eLearn 10 (2005)
3. Advanced Distributed Learning (ADL). Sharable Content Object Reference Model (SCORM) (2004), `http://www.adlnet.org`
4. Rogers, C.P., Liddle, S.W., Chan, P., Doxey, A., Isom, B.: Web 2.0 learning platform: Harnessing collective intelligence. Online Submission 8, 16-33 (2007), `http://www.eric.ed.gov/ERICWebPortal/detail?accno=ED498811`

5. Rey López, M., Díaz Redondo, R., Fernández Vilas, A., Pazos Arias, J., López Nores, M., García Duque, J., Gil Solla, A., Ramos Cabrer, M.: T-MAESTRO and its Authoring Tool: Using Adaptation to Integrate Entertainment into Personalized T-learning. Multimedia Tools and Applications 40(3) (2008)
6. Fernández Vilas, J., Pazos Arias, J., Fernández Vilas, A.: Virtual Web Services: An Extension Architecture to Alleviate Open Problems in Web Services Technology. IGI Global (2008)
7. Fernández Vilas, J., Pazos Arias, J., Fernández Vilas, A.: VWS: Applying Virtualization Techniques to Web Services. International Journal of Computer Science and Network Security 6(5B) (2006)
8. Rey López, M., Díaz Redondo, R., Fernández Vilas, A., Pazos Arias, J.: T-learning 2.0: a personalized hybrid approach based on ontologies and folksonomies. In: Computational Intelligence for Technology Enhanced Learning. Springer (2010)
9. Curbera, F., Khalaf, R., Mukhi, N., Tai, S., Weerawarana, S.: The next step in Web services. Communications of the ACM 46 (2003)
10. Vossen, G., Westerkamp, P.: Why service-orientation could make e-learning standards obsolete. International Journal of Technology Enhanced. Learning 1(1/2) (2008)
11. Dong, B., Zheng, Q., Yang, J., Li, H., Qiao, M.: An E-learning Ecosystem Based on Cloud Computing Infrastructure. In: 9th IEEE International Conference on Advanced Learning Technologies (2009)
12. Dong, B., Zheng, Q., Yang, J., Li, H., Qiao, M.: An. BlueSky Cloud Framework: An E-Learning Framework Embracing Cloud Computing. In: 1st International Conference on Cloud Computing (2009)
13. Sulistio, A., Reich, C., Dölitzscher, F.: Cloud Infrastructure & Applications. In: Proceedings of the 1st International Conference on Cloud Computing (2009)
14. Vouk, M., Averitt, S., Bugaev, M., Kurth, A., Peeler, A., Shaffer, H., Sills, E., Stein, S., Thompson, J.: Powered by VCL - Using Virtual Computing Laboratory (VCL). In: Proc. 2nd International Conference on Virtual Computing (2008)
15. Rey López, M., Díaz Redondo, R., Fernández Vilas, A., Pazos Arias, J.: Use of Folksonomies in the Creation of Learning Experiences for Television. Upgrade. Monograph: Technology-Enhanced Learning IX(3) (2008)
16. Hotho, A., Jäschke, R., Schmitz, C., Stumme, G.: Information Retrieval in Folksonomies: Search and Ranking. The Semantic Web: Research and Applications 4011, 411–426 (2006)

Part VI
Conclusions and Open Trends

Chapter 11
Conclusiones and Open Trends

José J. Pazos Arias, Ana Fernández Vilas, and Rebeca P. Díaz Redondo

Abstract. This book covers new advances on recommender systems for the Social Web. The contributed chapters look through general aspects related to recommenders, their legal effects, the problem of interoperability and the social influence for recommendation (trust and groups). Finally, two differentes applications of social recommendation are also shown. This chapter include the authors' view about the open trends and the future of recommendation. Specifically, we understand the current research lines might be ascribed to the following areas (i) the application of data-mining and SNA (Social Network Analysis) to obtain a Social Aggregation Model; (ii) the need for integrating different models and data structures to make interoperability in the Social Web feasible; and (iii) no matter how recommender systems will improve their precision and recall, its deployment must face with and find their natural arrangement in cloud computing environments.

11.1 Introduction

The scope of the recommendation of both content and services cannot be considered newly born, although its application may be considered in full swing. Social networking sites and the translation of personal relationships to the Web using new technologies are currently highly topical, as well. Precisely, in the junction of these two pillars, content recommendations and social relations, is based this book, which covers interesting aspects ranging from legal issues related to those recommenders in the Web 2.0 to specific cases of implementation of these mechanisms.

In the first part of the book, the more general aspects related to recommenders based on social relationships are condensed. In particular, the first chapter provides an overview of the state of the art in this field comparing collaborative filtering and

José J. Pazos Arias · Ana Fernández Vilas · Rebeca P. Díaz Redondo
Department of Telematic Engineering
University of Vigo
e-mail: {jose,avilas,rebeca}@det.uvigo.es

J.J. Pazos Arias et al.: Recommender Systems for the Social Web, ISRL 32, pp. 211–222.
springerlink.com

recommendation techniques, more traditional, with new techniques based on social relationships. The second chapter, on the other hand, deals with legal issues related to the acquisition of data and behavioral patterns of subscribers, which are collected with the intention of being used in these new strategies of recommendation, this chapter especially stresses the consequences that decentralization has and the importance of the explicit consent of users to collect and use their data.

The second part of the book is devoted to the analysis of interoperability in the field of recommendation. Also composed of two chapters. In the first one the authors study the challenges of tag recommendation systems, analyzing in depth the state of the art in the field and evaluating various approaches published in the literature to offer a broad view of the related problems. The second chapter is devoted to an aspect of particular interest to recommendation engines, how to select the most suitable ontology to support a process of recommendation for a specific scope. This article establishes a set of metrics for assessing the suitability of a specific ontology according to its wealth, its affinity with the field goal, the evaluation of its users, and so on.

The third part is dedicated to the study of trust and how this is essential in any process of recommendation involving group of users that maintain certain sort of relationship or social bond. The first chapter analyzes how it is feasible to use the information obtained from the implicit social network to improve the collaborative recommendation processes through the creation of dynamic and suitable neighborhoods for each specific target. The second one addresses the problem of inferring trust relationships between users that do not offer any explicit action. With this aim, the authors propose a strategy that obtains semantic relationships of trust and reputation, transparently to the user, from the results in previous recommendations.

The analysis of recommendation performed for groups is the core of the fourth part of the book. The first of its chapters is dedicated to providing a comparative study of the different methods of content recommendations for groups that usually stand out in the literature, ending with a set of open research fronts in this area, which are directly related to the field of social recommendation. The second chapter incorporates the concepts of friendship and membership to the field of collaborative recommendation based on social relationships, enriching the latter and providing more accurate results.

Finally the book ends with a part showing two different applications of the concepts previously treated. Specifically in the first chapter it is discussed how to recommend content for mobile devices, for which users' geographic location is an essential element. Being more specific, it is described a prediction system which sustains tours tourism content recommendation for mobile phone users. The last chapter, however, addresses an entirely different scope: educational. In this chapter the authors propose to change the traditional view of a the standard of education and support the recommendation of learning content, e-learning tools and educational plans based on the opinions and judgments that users make, which are strongly conditioned by the bonds of trust and reputation among them.

Of course, the chapters gathered here offer an overview of the problems linked to the content recommendation taking into account the relationships between users, but obviously also open a number of fields of study, whose detailed analysis would allow to obtain filtering and recommendation systems with more accurate results and more acceptation by their users. Specifically, we understand the current research lines might be ascribed to the following areas. Firstly, how should be taken into account the influence of group of users and trust or reputation relationships among them to improve the recommendation process? In this sense, social networks analysis is essential to extract important information that subscribers keep in the social site. Several approaches try to apply data-mining techniques, obviously taking into consideration the security restrictions the social sites impose. Secondly, how to structure the available information about users and content to facilitate the recommendation tasks? How to make feasible the integration of different data structures to support the characterization of both content and users? Currently, this open issue is devoted to reconcile collaborative tagging with semantic reasoning. And finally, how far will recommender systems are going to be conditioned by the characteristics of cloud computing? The following sections describe in more detail the lines on which the researchers are currently working in each one of these aspects.

11.2 Social Aggregation Model: Groups and Trust in the Network

Social networks are characterized by their subscribers can actively expose their interests and personal data (which together constitutes their profile). From the social network sites they are encouraged to seek out other subscribers and identify them as "friends" (the most usual term), creating their own "family" or "friend group". Finally, the social network sites set the necessary mechanisms to make it easy the interaction and exchange of information among members of a friend group and between them and other subscribers of the network [8].

Extracting information about the relationships between social networks' subscribers and the interaction supported by this kind of structures, which is usually known as social network analysis, is becoming one of the strongest areas where experts are currently working. Although there are different approaches, all of them assume modeling and representing the social network by using mathematical tools such as graphs or matrix structures. Based on this representation, traditional aspects that have arisen in the sociology [62] field are now applied in this technological context, such as connectivity, density, distance, diameter, centrality, and so on. Each of these concepts has a correspondent issue in the usual social relationships (like leadership, privacy, reputation, emotional contagion, etc.). In the literature there are several results [23, 47, 31] showing that social network analysis is an essential tool to obtain information of interest that allow supporting specific services (for example, recommenders) to its subscribers.

If there are two concepts that became central to the popularity of social networking these are surely the explicit treatment of trust and groups of users around an interest, problem or common cause. Both concepts should be reflected in the recommendation processes linked to social networks and, in fact, both are added value elements in the recommendation. Especially, if we combine these priceless sources and the advances on social network analysis in order to enhance recommendation in these contexts. If recommender systems have proven their key role in facilitating the user access to resources, where sharing resources has become social (in the context of social media), it is natural for recommendation strategies using the network structure to improve the precision and relevance of their results. Social recommender systems take into account trust-based relationships to accumulate users' opinions and calculate their predictions. The treatment of trust in social networking sites is, however, rather simplistic, users assign a trust value describing its connection with other users, without any explicit context or history which "explains" that trust. The term trust has been used in the field of multi-agent systems, where Gambetta [22] defines trust (or, symmetrically, distrust) as "a particular level of the subjective probability with which an agent assesses that another agent or group of agents will perform a particular action". In the context of social media, perhaps a more appropriate definition is the one by Sztompka [58]; "Trust is a bet about the future contingent actions of other", a definition that combines both "belief" (in how will be the future behavior of another person) and "commitment" to act on that belief. On the other hand, although the concept of reputation is closely linked to the concept of trust, its nature and dynamics are clearly different. According to the Oxford dictionary definition "Reputation is what is generally said or believed about a person's or thing's character or standing", and this is precisely the definition that corresponds with the notion of reputation in social networks [21, 41]. We face a measure that is visible to all social network members and derived from the underlying social structure.

We can find in the literature a considerable number of papers which address the use of models of reputation and trust in recommender systems [38, 29, 60, 3]. The inclusion of these models should consider different aspects: propagation of trust and distrust, relative value of similarity and social influence (trust) between two users, the inference of reputation from the explicit trust, and even the dynamic management of trust. Undoubtedly, one of the key aspects in the management of trust in social networks, for recommendation or for other purposes, is how to infer the level of trust between users not directly connected in the network, some proposals are the Advogato's metric [37], the Appleseed algorithm [66] and ones proposed by Guha [26] and Kuter [32]. Once we managed to obtain a measure of trust between users, in a context of collective intelligence, it is natural to use the opinion and assessment of "neighbors" to the target user on a strategy of collaborative filtering. In this case, the problem is to determine a neighborhood that properly weighs both the profile's likeness and the trust between users, i.e. establishing the right balance between similarity and social influence [17]. We can find outstanding proposals in this area in the following jobs [33, 16, 64, 15, 39]. Finally, one of the open lines that have come up in this area, which flows into the area of data-mining, is that the information inferred from the analysis of these networks will unveil reputation and trust relationships, such as managing these relationships in a dynamic way, and even will allow us to predict future relationships between subscribers, which obviously might facilitate the work of the recommendation engines.

On the other hand, as mentioned, the grouping of users behind a cause, issue or common interest is another defining aspect of the social Web in general and of social networking sites specifically. If the availability of trust information can improve the recommendation process, the existence of groups of users motivates the need for group recommendation strategies, i.e. strategies which, driven either by similarity or trust measures, predict whether a resource is appealing to a group of users. Regarding the definition of such strategies, a crucial aspect to consider is the degree of heterogeneity inside the group.

Many current approaches, in which individual profiles are available, opt for merging those profiles to build a virtual representative for the group, to which existing individualized algorithms can be applied [30, 51, 63, 9, 10]. This approach poses serious doubts when there is considerable heterogeneity among group members [19, 43], since opposing characteristics that end up canceling each other out, resulting in a significant loss of knowledge. Under these circumstances, several works have studied the generation of personal recommendations followed by a process which merges suggestions in order to maximize the satisfaction of the group [42, 44, 34, 2, 18]. Therefore, although group recommendation has distinguishing features that make it worthy of a theoretical corpus itself, very often rely on predictions that the traditional recommendation algorithms (i.e. individual) made for each of the members of the group separately.

11.3 Reconciling Collaborative Tagging with Semantic Reasoning

The achievement of the objectives pursued by the growing wave of personalization in the social Web needs to capture both the characteristics of resources in the Web and the characteristics of its users (use frequency, ratings, opinions, context, etc.). In line with this, the relentless growth in user participation has led to a democratization of resources on the Web, not only regarding to consumption but also to resources production. It is because of this avalanche of prosumers (which reflects the dual role of users: producers-consumers) from where collaborative tagging and the folksonomy concept come from.

The process by which users create and manage collaboratively the labels (tags) linked to available resources such as videos, photos, web pages, etc., in order to describe them is called collaborative tagging. In collaborative tagging systems, not only users may create and tag new resources but also may categorize and tag resources added by others [24]; Del.icio.us, and Flickr are examples of this approach. VanderWal Thomas coined the term folksonomy in 2004 [59], "the user-created bottom-up categorical structure an emergent Development with thesaurus" to refer to structures that are generated using this method of classification. This is an opposite approach to the Semantic Web philosophy, which proposes a top-down approach where users use a previously agreed vocabulary to describe their resources, facilitating the interoperability on the Web. Unfortunately, it is not obvious how it can be got that users describe the resources based on a consensual domain model:

more or less complex, ontology-based or folksonomy-based or any other structure. On the other hand, the Web has actually shown that users, at least the most active ones, tag resources using short text descriptions or using any label, which enables a folksonomy-based approach, i.e. a bottom-up approach in which assigning any kind of semantics is delayed.

The controversy between using a controlled vocabulary, quasi-static and centrally produced, versus using a free-text, dynamic and distributedly produced, constitutes an unnecessary barrier between Semantic Web and the social Web. The former claims a semantic infrastructure based on classification mechanisms using a defined relationship among terms (usually a taxonomy or ontology) with the aim of reasoning on the Web, automatizing the Web capabilities or improving them. Orthogonally to this, the strategic contribution of the social Web enables users to be part in a "collective intelligence" through collaborative tagging (usually in a folksonomy). However, collaborative tagging (using folksonomies) and formal tagging (using ontologies) are not opposite approaches, but complementary. In fact, the adequate combination of both of them would retain the potential of semantic reasoning, overcoming its practical and social barriers.

Characterizing users and content is essential to addressing the recommendation process and indeed, the model used to support these descriptions (ontology, taxonomy, folksonomy, hybrid) determines to a large extent, the algorithmic design and, consequently, the potential and performance of recommendation engine. In any case, assessing the semantic similarity between pairs (user-content, content-content, user-user) is the basis for this recommendation engine.

In the literature different solutions are proposed. For instance, [20] supports the calculation of similarity through ontological structures, and the same approach is applied in very different fields such as electronic commerce [48, 36], news [13, 11, 40], scientific articles [45] or the contents of TV [7]. Despite its many benefits, and while interoperability is not a reality in the Semantic Web, recommender systems based on ontologies, including taxonomies, are limited by the need to operate in local and controlled environments. On the other hand, even with a loose structure that undoubtedly leads to a lower potential for reasoning, folksonomies are also being used to achieve the same goals, essentially an assessment of similarity between terms that support the process of recommendation.

The literature also includes approaches in this regard, as is the case of FolkRank algorithm [28], the metric of similarity between labels in [56] and the approach based on commonsense reasoning in [49]. Specifically, in the case of content recommendation, [50] uses the information in a social bookmarking system for recommending new websites; [57] combines an ontology (created from the information in the video NetFix) and the Internet Movie DataBase folksonomy for recommending movies; [52] uses an unstructured model, folksonomy, in order to accommodate the large amount of information on Sensor Web; and finally [54], where information that represents the interests of the user is extracted from a folksonomy.

Although there are static approaches to combine ontological and folksonomic structures, those methodologies that try to make explicit the semantics behind the social tagging are the most remarkable ones; which some authors call semantic-enrichment [4]. In these approaches there are some mechanisms, like cleaning processes, analysis of co-occurrence, clustering, classification and extraction of semantic relationships, which are used to support the coexistence between folksonomies and ontologies where the former provide a simple mechanism for tagging and this information is automatically translated into the latter, hiding the underlying complexity to users and setting up the idea of emergent semantics [1] (a process that uses the interaction between users and machines to infer meaning). Moreover, the search for semantic and social recommendation strategies should entail a step further in reasoning over folksonomic structures and how integrated them with other more settled reasoning techniques in the semantic Web field, which has been called Social Semantic Web [46].

Despite recent advances in similarity metrics that support the recommendation, both in the case of ontological models as in the case of folksonomic models, there is not a body of techniques that support the hybrid modeling of users and products, so that collaborative tagging and ontological characterization can naturally coexist in the Social Web. Beyond coexistence, the objective is the integrated management of both forms of description, whose origin and structure is radically different. There are several factors that hinder this integration, without going any further, the difficulty of comparing the results obtained with both models with each other, either through simulations or through pilot projects. But this is an essential field of research to achieve better recommendation engines.

11.4 Recommending in the Cloud

The Social Web has erupted accompanied by a growing wave of decentralization at all levels of Web value chain: content production, service provision, decision making, tagging, etc. Recommendation systems are not immune to this rapidly rising tendency. Although recommender systems traditionally have been operating in centralized client-server environments, providing services to a closed community of users, there is a growing trend in academia that comes to reconciling these recommendation systems with P2P-related approaches. This decentralization trend reaches its highest expression in the main idea underlying cloud computing, the conception of the Web as a cloud of services, resources and infrastructures fully interoperable. That is, a new way of providing computing resources based on the virtualization of storage, computing power, development platforms and applications. In this context, virtualization is intended to facilitate the use of resources, reduce costs associated with its management, or guarantee immediate scalability on them, thereby improving their profitability. This reinvention of the Web provides new opportunities for recommender systems, but also puts new challenges.

First, the new scenario reaches global levels. It is not to recommend an item from a closed catalogue of products, but rather, for example, to recommend a TV show among those offered by all IPTV providers in the cloud. This new scenario adds a new ingredient to the classic problem of recommendation: the service provider and consequently their characteristics. The recommendation algorithm should not only be able to find the resource (content, service, etc..) which is more suitable for the user, but also be able to identify a provider in the cloud that is appealing to the user. Thus, taking into account other components of the problem, parameters such as quality, reliability and dependability of the supplier move into the foreground [35, 2].

On the other hand, cloud computing seems a suitable way to address the problem of scaling in recommender systems, especially in collaborative filtering. When the emphasis shifts from mass customers to niche customers, there is not a small set of prototype-users or clusters so that scaling problems are exacerbated. In this regard various studies have proposed solutions to distribute the computation of the recommendation [53, 27, 14, 25] and, more recently, approaches over cloud platforms like Apache Hadoop^{TM}[65].

A model for cloud recommendation would alleviate not only the scale problems in collaborative filtering, but also those that are consequence of implementing the recommendation in closed or isolated environments, those linked to the spread of the data, and those that result from the emergence of new users and new product in the system [5]. The problem of data acquisition, which is inherent in traditional recommender systems, can be alleviated through a model which virtualizes the computation of the predictions, i.e. a model which orchestrates a set of third-party recommendation services in the cloud; services which may operate according to different recommendation strategies, and may reason on the Web of Data [6] through the best practices of Linked Data [1].

Apart from the global nature of the data, changes in Web habits linked to cloud computing out the possibility of having a centralized architecture in which data is stored about users' consumption, tastes, assessments, etc. Deployment solutions should focus on models where both user preferences (profiles) and information that describes the products (metadata) are owned by users who access the resources in the cloud, final beneficiaries of the recommendation, and never owned by the recommendation system. This user-oriented conception (as opposed to product-oriented) also comes to the aid of growing concern in the privacy of data, since users' opinions and habits are maintained in structures local to the user, but possibly located in the cloud [12, 55].

From the viewpoint of cloud user, it is not desirable a closed model where the user logs into a system and get recommendations for items on that system. Users will choose one or another recommendation service according to their preferences and/or needs. Even more, using a recommendation service or another will change over time and, regardless of the business model behind it, the recommendation service may be unable to update the user profile based on the history records. The user may never have used such a recommendation system, and also possibly not use the system again [61]. Again, the Web of Data is critical to the feasibility of such scenario

[1] `http://www.w3.org/DesignIssues/LinkedData.html`

since recommendations services may not use information managed by them and will import information about users and resources from external sources. Whereas the extraction of such information must address problems related to privacy and security; the integration and, especially, the identification of relevant information requires significant advances in semantic interoperability of metadata.

In summary, in regard to the unstoppable advance of cloud computing, there are two key challenges. On the one hand, refactoring the traditional architectonic model for recommendation into a cloud model; and, on the other hand, the conception of a recommendation model that is independent of the products to recommend and therefore orthogonal to the services specific.

Acknowledgements. This work has been partially funded by the Ministerio de Educación y Ciencia (Gobierno de España) research project TIN2010-20797 (partly financed with FEDER funds), and by the Consellería de Educación e Ordenación Universitaria (Xunta de Galicia) incentives file CN 2011/023 (partly financed with FEDER funds).

References

1. Aberer, K., Cudré-Mauroux, P., Catarci, A.M.O.T., Hacid, M.-S., Illarramendi, A., Kashyap, V., Mecella, M., Mena, E., Neuhold, E.J., de Troyer, O., Risse, T., Scannapieco, M., Saltor, F., de Santis, L., Spaccapietra, S., Staab, S., Studer, R.: Emergent semantics principles and issues. In: Lee, Y., Li, J., Whang, K.-Y., Lee, D. (eds.) DASFAA 2004. LNCS, vol. 2973, pp. 25–38. Springer, Heidelberg (2004)
2. Alhamad, M., Dillon, T., Chang, E.: Sla-based trust model for cloud computing. In: International Conference on Network-Based Information Systems (2010)
3. Andersen, R., Borgs, C., Chayes, J., Feige, U., Flaxman, A., Kalai, A., Mirrokni, V., Tennenholtz, M.: Trust-based recommendation systems: an axiomatic approach. In: Proceeding of the 17th International Conference on World Wide Web (2008)
4. Angeletou, S., Sabou, M., Specia, L., Motta, E.: Bridging the gap between folksonomies and the semantic web: An experience report. In: SemNet (2007)
5. Heitmann, B., Hayes, C.: Using linked data to build open, collaborative recommender systems. In: AAAI Spring Symposium: Linked Data Meets Artificial Intelligence (2007)
6. Bizer, C., Heath, T., Berners-Lee, T.: Linked data - the story so far. International Journal on Semantic Web and Information Systems (2009)
7. Blanco-Fernández, Y., Pazos-Arias, J.J., Gil-Solla, A., Ramos-Cabrer, M., López-Nores, M., García-Duque, J., Fernández-Vilas, A., Díaz-Redondo, R.P., Bermejo-Muñoz, J.: A flexible semantic inference methodology to reason about user preferences in knowledge-based recommender systems. Knowlegde-Based Systems 21(4), 305–320 (2008)
8. Boyd, D.M., Ellison, N.B.: Social network sites: Definition, history, and scholarship. Journal of Computer-Mediated Communication 13(1) (2007)
9. Cantador, I., Castells, P.: Building emergent social networks and group profiles by semantic user preference clustering (2006)
10. Cantador, I., Castells, P., Vallet, D.: Enriching group profiles with ontologies for knowledge-driven collaborative content retrieval. In: 15th IEEE International Workshops on Enabling Technologies: Infrastructures for Collaborative Enterprises (2006)
11. Cantador, I., Bellogín, A., Castells, P.: Ontology-based personalised and context-aware recommendations of news items. In: Proceedings of the 2008 IEEE/WIC/ACM International Conference on Web Intelligence and Intelligent Agent Technology (2008)

12. Castagnos, S., Boyer, A.: Modeling Preferences in a Distributed Recommender System. In: Conati, C., McCoy, K., Paliouras, G. (eds.) UM 2007. LNCS (LNAI), vol. 4511, pp. 400–404. Springer, Heidelberg (2007)
13. Castells, P., Fernandez, M., Vallet, D.: An adaptation of the vector-space model for ontology-based information retrieval. IEEE Transactions on Knowledge and Data Engineering 19(2), 261–272 (2007)
14. Chen, M.-H., Lin, K.C.-J., Kung, C.-C., Chou, C.-F., Tu, C.-J.: On the design of the semantic p2p system for music recommendation. In: International Symposium on Parallel and Distributed Processing with Applications (2010)
15. Chen, S., Luo, T., Liu, W., Xu, Y.: Incorporating similarity and trust for collaborative filtering. In: Proccedings of Sixth International Conference on Fuzzy Systems and Knowledge Discovery (2009)
16. Chen, Z.M., Jiang, Y., Zhao, Y.: A collaborative filtering recommendation algorithm based on user interest change and trust evaluation. JDCTA: International Journal of Digital Content Technology and its Applications 4 (2010)
17. Crandall, D., Cosley, D., Huttenlocher, D., Kleinberg, J., Suri, S.: Feedback effects between similarity and social influence in online communities. In: Proceeding of the 14th ACM SIGKDD International Conference on Knowledge Discovery and Data Mining (2008)
18. Das, A., Datar, M., Garg, A., Rajaram, S.: Google news personalization: scalable online collaborative filtering. In: Proceedings of the 16th International Conference on World Wide Web (2007)
19. de Campos, L.M., Fernandez-Luna, J.M., Huete, J.F., Rueda-Morales, M.A.: Group recommending: A methodological approach based on bayesian networks. In: IEEE Intl. Conf. on Data Engineering Workshop (2007)
20. Euzenat, J., Shvaiko, P.: Ontology matching. Springer, New York (2007)
21. Freeman, L.C.: Centrality on social networks. Social Networks 1, 215–239 (1979)
22. Gambetta, D.: Can We Trust Trust? In: Trust: Making and Breaking Cooperative Relations, electronic edition, Department of Sociology, University of Oxford (2000)
23. Garton, L., Haythornthwaite, C., Wellman, B.: Studying online social networks. Journal of Computer-Mediated Communication 3(1) (1997)
24. Golder, S.A., Huberman, B.A.: The structure of collaborative tagging systems. Journal of Information Science, 198–208 (2006)
25. Gong, S., Ye, H., Su, P.: A peer-to-peer based distributed collaborative filtering architecture. In: International Joint Conference on Artificial Intelligence (2009)
26. Guha, R., Kumar, R., Raghavan, P., Tomkins, A.: Propagation of trust and distrust. In: Proceeding of the International Conference on World Wide Web (2004)
27. Han, P., Xie, B., Yang, F., Shen, R.: A scalable p2p recommender system based on distributed collaborative filtering. Expert Systems With Applications 7(2) (2004)
28. Hotho, A., Jäschke, R., Schmitz, C., Stumme, G.: Information Retrieval in Folksonomies: Search and ranking. In: Sure, Y., Domingue, J. (eds.) ESWC 2006. LNCS, vol. 4011, pp. 411–426. Springer, Heidelberg (2006)
29. Jamali, M., Ester, M.: Trustwalker: a random walk model for combining trust-based and item-based recommendation. In: Proceedings of the 15th ACM SIGKDD International Conference on Knowledge Discovery and Data Mining (2009)
30. Jameson, A.: More than the sum of its members: Challenges for group recommender systems. In: Proceedings of the Int'l Working Conference on Advanced Visual Interfaces (2004)
31. Kumar, R., Novak, J., Tomkins, A.: Structure and Evolution of Online Social Networks. In: Link Mining: Models, Algorithms, and Applications. Springer (2010)

32. Kuter, U., Golbeck, J.: Using probabilistic confidence models for trust inference in web-based social networks. ACM Transactions on Internet Technology (2010)
33. Lathia, N., Hailes, S., Capra, L.: Trust-based collaborative filtering. In: IFIPTM 2008. IFIP AICT (2008)
34. Lee, S., Lee, D., Lee, S.: Personalized dtv program recommendation system under a cloud computing environment. IEEE Transactions on Consumer Electronics 56(2) (2010)
35. Lee, S., Lee, D., Lee, S.: Personalized dtv program recommendation system under a cloud computing environment. IEEE Transactions on Consumer Electronics 56(2), 1034–1042 (2010)
36. Lee, T., Chun, J., Shim, J., Lee, S.-G.: An ontology-based product recommender system for b2b marketplaces. International Journal of Electronic Commerce 11(2), 125–155 (2007)
37. Levien, R.: Attack Resistant Trust Metrics. PhD thesis, UC Berkeley (2004)
38. Li, Y.-M., Kao, C.-P.: Trepps: A trust-based recommender system for peer production services. Expert Systems with Applications 36(2) (2009)
39. Ma, H., King, I., Lyu, M.R.: Learning to recommend with social trust ensemble. In: Proceedings of the 32nd International ACM SIGIR Conference on Research and Development in Information Retrieval (2009)
40. Maidel, V., Shoval, P., Shapira, B., Taieb-Maimon, M.: Evaluation of an ontology-content based filtering method for a personalized newspaper. In: Proceedings of the 2008 ACM Conference on Recommender Systems (2008)
41. Marsden, P.V., Lin, N. (eds.): Social Structure and Network Analysis. Sage Publications, Beverly Hills (1982)
42. Masthoff, J.: Modeling a group of television viewers. In: Future TV: Adaptive Instruction in Your Living Room Workshop
43. Masthoff, J.: Group modeling: Selecting a sequence of television items to suit a group of viewers. User Modeling and User- Adaption Interaction 14, 37–85 (2004)
44. Masthoff, J., Gatt, A.: In pursuit of satisfaction and the prevention of embarrassment: affective state in group recommender systems. User Modeling and User- Adaption Interaction 16, 281–319 (2006)
45. Middleton, S.E., Shadbolt, N.R., de Roure, D.C.: Ontological user profiling in recommender systems. ACM Transactions on Information Systems 22(1), 54–88 (2004)
46. Mikroyannidis, A.: Towards a social semantic web. Computer 40(11), 113–115 (2007)
47. Mislove, A., Marcon, M., Gummadi, K.P., Druschel, P., Bhattacharjee, B.: Measurement and analysis of online social networks. In: Proceedings of the 7th ACM SIGCOMM Conference on Internet Measurement (2007)
48. Moosavi, S., Nematbakhsh, M., Farsani, H.K.: A semantic complement to enhance electronic market. Expert Systems with Applications 36, 5768–5774 (2009)
49. Nauman, M., Hussain, F.: Common sense and folksonomy: Engineering an intelligent search system. In: International Conference on Information and Emerging Technologies (2007)
50. Niwa, S., Doi, T., Honiden, S.: Web page recommender system based on folksonomy mining. In: Information Technology: New Generations (2006)
51. Pizzutilo, S., De Carolis, B., Cozzolongo, G., Ambruoso, F.: Group modeling in a public space: methods, techniques, experiences. In: Procs of the 5th WSEAS International Conference on Applied Informatics and Communications (2005)
52. Rezel, R., Liang, S., Kim, K.-S.: A folksonomy-based recommendation system for the sensor web. In: Tanaka, K., Fröhlich, P., Kim, K.-S. (eds.) W2GIS 2011. LNCS, vol. 6574, pp. 64–77. Springer, Heidelberg (2010)

53. Das, A.S., Datar, M., Garg, A., Rajaram.:Google news personalization: scalable online collaborative filtering. In: Proceedings of the 16th International Conference on World Wide Web (2007)
54. Saito, J., Yukawa, T.: Extracting user interest for user recommendation based on folksonomy. IEEE Transactions on Information and Systems (2011)
55. Shokri, R., Pedarsani, P., Theodorakopoulos, G., Hubaux, J.-P.: Preserving privacy in collaborative filtering through distributed aggregation of offline profiles. In: Proceedings of the Third ACM Conference on Recommender Systems (2009)
56. Specia, L., Motta, E.: Integrating Folksonomies with the Semantic Web. In: Franconi, E., Kifer, M., May, W. (eds.) ESWC 2007. LNCS, vol. 4519, pp. 624–639. Springer, Heidelberg (2007)
57. Szomszor, M., Cattuto, C., Alani, H., O'Hara, K., Baldassarri, A., Loreto, V., Servedio, V.D.: Folksonomies, the semantic web, and movie recommendation. In: 4th European Semantic Web Conference (2007)
58. Sztompka, P.: Trust: A Sociological Theory. Cambridge University Press (1999)
59. Vander Wal, T.: Folksonomy coinage and definition (2007), `http://www.vanderwal.net/folksonomy.html`
60. Walter, F., Battiston, S., Schweitzer, F.: A model of a trust-based recommendation system on a social network. Autonomous Agents and Multi-Agent Systems 16(1) (2008)
61. Wang, S., Xie, Y., Fang, M.: A collaborative filtering recommendation algorithm based on item and cloud model. Wuhan University Journal of Natural Sciences 16(1) (2011)
62. Wasserman, S., Faust, K.: Social network analysis: Methods and applications
63. Yu, Z., Zhou, X., Hao, Y., Gu, J.: Tv program recommendation for multiple viewers based on user profile merging. User Modeling and User-Adapted Interaction 16(1), 63–82 (2006)
64. Zhang, F., Bai, L., Gao, F.: A User Trust-Based Collaborative Filtering Recommendation Algorithm. In: Qing, S., Mitchell, C.J., Wang, G. (eds.) ICICS 2009. LNCS, vol. 5927, pp. 411–424. Springer, Heidelberg (2009)
65. Zhao, Z.-D.,, M.: s. Shang. User-based collaborative-filtering recommendation algorithms on hadoop. In: International Workshop on Knowledge Discovery and Data Mining (2010)
66. Ziegler, C.N.: Towards Decentralized Recommender Systems. PhD thesis, Albert-Ludwigs-Universit at Freiburg (2005)